EDITION

3

Student Solutions Manual for Taylor and Gilligan's
Applied Calculus

Jeanne Bowman
University of Cincinnati

Brooks/Cole
Publishing
Company
Pacific Grove,
California

Brooks/Cole Publishing Company
A Division of Wadsworth, Inc.

Printed in the United States of America

10 9 8 7 6 5 4 3 2 1

ISBN 0-534-19352-8

Sponsoring Editor: Audra C. Silverie
Editorial Assistant: Carol Ann Benedict
Production Coordinator: Dorothy Bell
Cover Design: Susan Haberkorn
Art Coordinator: Lisa Torri
Printing and Binding: Malloy Lithographing, Inc.

Preface

This manual is designed to accompany *Applied Calculus* by Claudia Taylor and Lawrence Gilligan. It contains complete solutions for all odd-numbered exercises in each section, plus solutions to all problems in the review exercises and chapter tests. Remember that many of the exercises can be solved correctly using methods other than the ones shown here.

I would like to thank Audra Silverie for her assistance in preparing this manual and Mary Ann Zuzow for her work in typing it.

Contents

CHAPTER 1: ALGEBRA REVIEW

Section 1.1 (pages 7-9)

1. $3x^2 - 7x + 3$

3. $2x - x^2 - 5 - 3x^2$
$= -4x^2 + 2x - 5$

5. $4x - 12 - 8x - 2$
$= -4x - 14$

7. $-2[5 - 3x + 6]$
$= -2[-3x + 11]$
$= 6x - 22$

9. $2x^4 - x^3 + 4x^2$

11. $2x^2 - 14x + 3x - 21$
$= 2x^2 - 11x - 21$

13. $5x - x[2x - x + 3]$
$= 5x - x[x + 3]$
$= 5x - x^2 - 3x$
$= 2x - x^2$

15. $2(-2)^3 - 3(-2) + 5$
$= 2(-8) - 3(-2) + 5$
$= -16 + 6 + 5$
$= -5$

17. $3(-2)^4 - (-2) + 6$
$= 3(16) - (-2) + 6$
$= 48 + 2 + 6$
$= 56$

19. $5(-2)^2(3) - 3(-2)(3)^2$
$= 5(4)(3) - 3(-2)(9)$
$= 60 + 54$
$= 114$

21. $(-2)^2 + (3)^2$
$= 4 + 9$
$= 13$

23. $\sqrt[4]{x^3}$

25. $4\sqrt{x}$

27. $\dfrac{3}{(x+2)^{1/2}}$

$= \dfrac{3}{\sqrt{x+2}}$

29. $x^{1/3}$

31. $-2x \cdot x^{1/4}$
$=-2x^{5/4}$

33. $\dfrac{2x}{x^{1/3}}$

$=2x^{2/3}$

35. $16^{1/2}$
$=\sqrt{16}$
$=4$

37. $(-27)^{1/3}$
$=\sqrt[3]{-27}$
$=-3$

39. $(-64)^{2/3}$
$=(\sqrt[3]{-64})^2$
$=(-4)^2$
$=16$

41. $5(16)^{3/4}$
$=5\left(\sqrt[4]{16}\right)^3$
$=5(2)^3$
$=5(8)$
$=40$

43. $\dfrac{-1}{\sqrt[3]{7-(-1)}}$

$=\dfrac{-1}{\sqrt[3]{8}}$

$=-\dfrac{1}{2}$

45. $\dfrac{5}{\sqrt{6}} \cdot \dfrac{\sqrt{6}}{\sqrt{6}}$

$=\dfrac{5\sqrt{6}}{\sqrt{36}}$

$=\dfrac{5\sqrt{6}}{6}$

47. $\dfrac{\sqrt{x}}{1+\sqrt{x}} \cdot \dfrac{1-\sqrt{x}}{1-\sqrt{x}}$

$=\dfrac{\sqrt{x}(1-\sqrt{x})}{1-\sqrt{x}+\sqrt{x}-x}$

$=\dfrac{\sqrt{x}-x}{1-x}$

49. $\dfrac{6}{\sqrt{x}-4} \cdot \dfrac{\sqrt{x}+4}{\sqrt{x}+4}$

$=\dfrac{6(\sqrt{x}+4)}{x+4\sqrt{x}-4\sqrt{x}-16}$

$=\dfrac{6\sqrt{x}+24}{x-16}$

51. $\dfrac{\sqrt{x}}{x} \cdot \dfrac{\sqrt{x}}{\sqrt{x}}$

$=\dfrac{\sqrt{x^2}}{x\sqrt{x}}$

$=\dfrac{x}{x\sqrt{x}}$

$=\dfrac{1}{\sqrt{x}}$

53. $\dfrac{\sqrt{x}+1}{2} \cdot \dfrac{\sqrt{x}-1}{\sqrt{x}-1}$

$= \dfrac{x - \sqrt{x} + \sqrt{x} - 1}{2(\sqrt{x}-1)}$

$= \dfrac{x-1}{2\sqrt{x}-2}$

55. $\dfrac{\sqrt{x}+\sqrt{x+1}}{2} \cdot \dfrac{\sqrt{x}-\sqrt{x+1}}{\sqrt{x}-\sqrt{x+1}}$

$= \dfrac{x - \sqrt{x}\sqrt{x+1} + \sqrt{x}\sqrt{x+1} - (x+1)}{2(\sqrt{x}-\sqrt{x+1})}$

$= \dfrac{x - (x+1)}{2\sqrt{x}-2\sqrt{x+1}}$

$= \dfrac{-1}{2\sqrt{x}-2\sqrt{x+1}}$

57. $x^{-2/3} \cdot x + x^{-2/3} \cdot 2x^{2/3}$

$= x^{1/3} + 2x^0$

$= x^{1/3} + 2$

59. $x^{1/2} \cdot x - x^{1/2} \cdot 1 + 2x^{-1/2} \cdot 3x$

$= x^{3/2} - x^{1/2} + 6x^{1/2}$

$= x^{3/2} + 5x^{1/2}$

61. $-3[4 - 4x + 2] + 4x$

$= -3[6 - 4x] + 4x$

$= -18 + 12x + 4x$

$= 16x - 18$

63. $2x \cdot x^2 + 2x \cdot x - 2x \cdot 1 + 3 \cdot x^2 + 3 \cdot x - 3$

$= 2x^3 + 2x^2 - 2x + 3x^2 + 3x - 3$

$= 2x^3 + 5x^2 + x - 3$

65. $\dfrac{2x}{(x^2+1)^{2/3}}$

$= 2x(x^2+1)^{-2/3}$

67. $(16)^{1/2} + (-27)^{1/3}$

$= \sqrt{16} + \sqrt[3]{-27}$

$= 4 + (-3)$

$= 1$

69.

$$\frac{2}{\sqrt{3}-\sqrt{5}}\cdot\frac{\sqrt{3}+\sqrt{5}}{\sqrt{3}+\sqrt{5}}$$

$$=\frac{2(\sqrt{3}+\sqrt{5})}{3+\sqrt{15}-\sqrt{15}-5}$$

$$=\frac{2(\sqrt{3}+\sqrt{5})}{3-5}$$

$$=\frac{2(\sqrt{3}+\sqrt{5})}{-2}$$

$$=-1(\sqrt{3}+\sqrt{5})$$

$$=-\sqrt{3}-\sqrt{5}$$

71. Total area: $x^2+9x+2x+18$

$$=x^2+11x+18$$

Section 1.2 (pages 14-15)

1. $2x^3(3x+1)$

3. $(x+3)(x+4)$

5. $(3x-2)(x+2)$

7. $(4-x)(2+x)$

9. $x^2(x^2-9)$

$$=x^2(x+3)(x-3)$$

11. x^3-3^3

$$=(x-3)(x^2+3x+9)$$

13. $3x(2x^2+x-1)$

$$=3x(2x-1)(x+1)$$

15. $(x-3)(5x+2)$

17. $2x^2(2x+1)^{-3}[-4x+3(2x+1)]$

$$=2x^2(2x+1)^{-3}[-4x+6x+3]$$

$$=2x^2(2x+1)^{-3}(2x+3)$$

$$=\frac{2x^2(2x+3)}{(2x+1)^3}$$

19. $(x+1)(x-4)^{-3}[-2(x+1)+3(x-4)]$

$=(x+1)(x-4)^{-3}[-2x-2+3x-12]$

$=(x+1)(x-4)^{-3}(x-14)$

$=\dfrac{(x-1)(x-14)}{(x-4)^3}$

21. $x^2(2x+1)^{-2/3}[3(2x+1)-x]$

$=x^2(2x+1)^{-2/3}[6x+3-x]$

$=x^2(2x+1)^{-2/3}(5x+3)$

$=\dfrac{x^2(5x+3)}{(2x+1)^{2/3}}$ or $\dfrac{x^2(5x+3)}{\sqrt[3]{(2x+1)^2}}$

23. $x^{-1/2}(1-x)[(1-x)-4x]$

$=x^{-1/2}(1-x)(1-5x)$

$=\dfrac{(1-x)(1-5x)}{x^{1/2}}$ or $\dfrac{(1-x)(1-5x)}{\sqrt{x}}$

$=\dfrac{(x-1)(5x-1)}{\sqrt{x}}$

25. $\dfrac{(x+3)(x-3)}{x-3}$

$=x+3$

27. $\dfrac{(x+3)(x-3)}{-1(x-3)}$

$=\dfrac{(x+3)}{-1}$

$=-x-3$

29. $\dfrac{x^2(x^4-16)}{x^4+5x^2+4}$

$=\dfrac{x^2(x^2+4)(x^2-4)}{(x^2+1)(x^2+4)}$

$=\dfrac{x^2(x^2-4)}{x^2+1}$

$=\dfrac{x^2(x+2)(x-2)}{x^2+1}$

31. $\dfrac{(x+1)(x-3)}{(x-3)^{1/2}}$

$=(x+1)(x-3)^{1/2}$ or $(x+1)\sqrt{x-3}$

33. $\dfrac{x(x+1)}{x^{1/2}}$

$= x^{1/2}(x+1)$ or $\sqrt{x}(x+1)$

35. $\dfrac{2 \cdot x}{x \cdot x} + \dfrac{1}{x^2}$

$= \dfrac{2x}{x^2} + \dfrac{1}{x^2}$

$= \dfrac{2x+1}{x^2}$

37. $\dfrac{x(x-3)}{(x+1)(x-3)} - \dfrac{2x(x+1)}{(x-3)(x+1)}$

$= \dfrac{x^2 - 3x}{(x+1)(x-3)} - \dfrac{2x^2 + 2x}{(x+1)(x-3)}$

$= \dfrac{x^2 - 3x - 2x^2 - 2x}{(x+1)(x-3)}$

$= \dfrac{-x^2 - 5x}{(x+1)(x-3)}$

39. $\dfrac{5y \cdot y^2}{1 \cdot y^2} - \dfrac{1}{y^2}$

$= \dfrac{5y^3}{y^2} - \dfrac{1}{y^2}$

$= \dfrac{5y^3 - 1}{y^2}$

41. $\dfrac{2(x-2)(x-3)}{(x-1)(x-2)(x-3)} + \dfrac{3(x-1)(x-3)}{(x-2)(x-1)(x-3)} + \dfrac{4(x-1)(x-2)}{(x-3)(x-1)(x-2)}$

$= \dfrac{2(x^2 - 5x + 6)}{(x-1)(x-2)(x-3)} + \dfrac{3(x^2 - 4x + 3)}{(x-1)(x-2)(x-3)} + \dfrac{4(x^2 - 3x + 2)}{(x-1)(x-2)(x-3)}$

$= \dfrac{2x^2 - 10x + 12}{(x-1)(x-2)(x-3)} + \dfrac{3x^2 - 12x + 9}{(x-1)(x-2)(x-3)} + \dfrac{4x^2 - 12x + 8}{(x-1)(x-2)(x-3)}$

$= \dfrac{9x^2 - 34x + 29}{(x-1)(x-2)(x-3)}$

43. $6x^3(3x-1)^{-3}[x + 2(3x-1)]$

$= 6x^3(3x-1)^{-3}[x + 6x - 2]$

$= 6x^3(3x-1)^{-3}(7x-2)$

$= \dfrac{6x^3(7x-2)}{(3x-1)^3}$

45. $x^2(2-x)^{-2/3}[4(2-x) - x]$

$= x^2(2-x)^{-2/3}[8 - 4x - x]$

$= x^2(2-x)^{-2/3}(8 - 5x)$

$= \dfrac{x^2(8-5x)}{(2-x)^{2/3}}$ or $\dfrac{x^2(8-5x)}{\sqrt[3]{(2-x)^2}}$

47. $(x+2)^2(x+4)^4[(x+4) + (x+2)]$

$= (x+2)^2(x+4)^4(2x+6)$

$= 2(x+3)(x+2)^2(x+4)^4$

49. $0.02x^{-3}(x^5 - 200)$

$= \dfrac{0.02(x^5 - 200)}{x^3}$

Section 1.3 (pages 22-23)

1. $4x + 1 = 3x - 2$

   $x = -3$

3. $5(x + 1) - 3 = 7$

   $5x + 5 - 3 = 7$

   $5x + 2 = 7$

   $5x = 5$

   $x = 1$

5. $\frac{1}{3}t - 2 = 1$

   $\frac{1}{3}t = 3$

   $t = 9$

7. $3(1 - y) + 2 = 0$

   $3 - 3y + 2 = 0$

   $-3y + 5 = 0$

   $-3y = -5$

   $y = \frac{5}{3}$

9. $x^2 - 6x = 0$

   $x(x - 6) = 0$

   $x = 0$ | $x - 6 = 0$

   | $x = 6$

11. $s^2 - s - 12 = 0$

    $(s - 4)(s + 3) = 0$

    $s - 4 = 0$ | $s + 3 = 0$

    $s = 4$ | $s = -3$

13. $2x^2 + 4x + 1 = 0$

    $a = 2,\ b = 4$ and $c = 1$

    $x = \dfrac{-4 \pm \sqrt{4^2 - 4(2)(1)}}{2(2)}$

    $x = \dfrac{-4 \pm \sqrt{16 - 8}}{4}$

    $x = \dfrac{-4 \pm \sqrt{8}}{4}$

    $x = \dfrac{-4 \pm 2\sqrt{2}}{4}$

    $x = \dfrac{2(-2 \pm \sqrt{2})}{4}$

    $x = \dfrac{-2 \pm \sqrt{2}}{2}$

15. $x^2 = 5$

    $\sqrt{x^2} = \pm\sqrt{5}$

    $x = \pm\sqrt{5}$

17. $3x^2 - 12 = 0$

    $3x^2 = 12$

    $x^2 = 4$

    $x = \pm 2$

19. $(x+5)^2 = 7$

$\sqrt{(x+5)^2} = \pm\sqrt{7}$

$x+5 = \pm\sqrt{7}$

$x = -5 \pm \sqrt{7}$

21. $t^2 + 2t - 2 = 0$

$a = 1, \ b = 2 \text{ and } c = -2$

$t = \dfrac{-2 \pm \sqrt{(2)^2 - 4(1)(-2)}}{2(1)}$

$t = \dfrac{-2 \pm \sqrt{4+8}}{2}$

$t = \dfrac{-2 \pm \sqrt{12}}{2}$

$t = \dfrac{-2 \pm 2\sqrt{3}}{2}$

$t = \dfrac{2(-1 \pm \sqrt{3})}{2}$

$t = -1 \pm \sqrt{3}$

23. $4x^2 - x + 2 = 0$

$a = 4, \ b = -1 \text{ and } c = 2$

$x = \dfrac{-(-1) \pm \sqrt{(-1)^2 - 4(4)(2)}}{2(4)}$

$x = \dfrac{1 \pm \sqrt{1 - 32}}{8}$

$x = \dfrac{1 \pm \sqrt{-31}}{8}$

no real solution

25. $x^4 - 5x^2 + 4 = 0$

$(x^2 - 4)(x^2 - 1) = 0$

$x^2 - 4 = 0$	$x^2 - 1 = 0$
$x^2 = 4$	$x^2 = 1$
$x = \pm 2$	$x = \pm 1$

27. $\sqrt[3]{x} = 4$

$(\sqrt[3]{x})^3 = 4^3$

$x = 64$

check: $\sqrt[3]{64} = 4$

29. $\sqrt[4]{1-x} = 0$

$(\sqrt[4]{1-x})^4 = 0^4$

$1 - x = 0$

$-x = -1$

$x = 1$

check: $\sqrt[4]{1-1} = 0$

31. $x^{3/2} - 8 = 0$

$$x^{3/2} = 8$$
$$(x^{3/2})^{2/3} = 8^{2/3}$$
$$x = (\sqrt[3]{8})^2$$
$$x = 2^2$$
$$x = 4$$

check: $4^{3/2} - 8 = 0$
$$(\sqrt{4})^3 - 8 = 0$$
$$8 - 8 = 0$$

33. $\dfrac{2}{\sqrt{x}} = 3$

$$2 = 3\sqrt{x}$$
$$\frac{2}{3} = \sqrt{x}$$
$$\left(\frac{2}{3}\right)^2 = (\sqrt{x})^2$$
$$\frac{4}{9} = x$$

check: $\dfrac{2}{\sqrt{\frac{4}{9}}} = 3$

$$\frac{2}{\frac{2}{3}} = 3$$
$$2 \cdot \frac{3}{2} = 3$$
$$3 = 3$$

35. $\dfrac{x+1}{x-3} = 2 \quad x \neq 3$

$$(x-3)\left(\frac{x+1}{x-3}\right) = (x-3)(2)$$
$$x + 1 = 2x - 6$$
$$7 = x$$

37. $\dfrac{x^2 - 4}{2x + 1} = 0$

$$x^2 - 4 = 0$$
$$x^2 = 4$$
$$x = \pm 2$$

39. $\dfrac{x - x^2}{\sqrt{x+3}} = 0$

$$x - x^2 = 0$$
$$x(1 - x) = 0$$

$x = 0$ | $1 - x = 0$
                $1 = x$

41. $x^2 - p = 0$

$$x^2 - (\sqrt{p})^2 = 0$$
$$(x + \sqrt{p})(x - \sqrt{p}) = 0$$

$x + \sqrt{p} = 0$ | $x - \sqrt{p} = 0$
$x = -\sqrt{p}$ | $x = \sqrt{p}$

43. $6x - x^2 - 5 = 0 \qquad 0 \leq x \leq 4$

$$0 = x^2 - 6x + 5$$
$$0 = (x - 1)(x - 5)$$

$x - 1 = 0$ | $x - 5 = 0$
$x = 1$ | $x = 5$

$x = 1$ since $0 \leq x \leq 4$

45.   (length)(width)(height)=volume

$$(x - 4)(x - 4)(2) = 968$$
$$(x^2 - 8x + 16)(2) = 968$$
$$x^2 - 8x + 16 = 484$$
$$x^2 - 8x - 468 = 0$$
$$(x - 26)(x + 18) = 0$$

$x - 26 = 0$ | $x + 18 = 0$
$x = 26$ | $x = -18$

The length and width should be 26 cm.

47.

$$-4t^2 + 76t + 80 = 360$$

$$-4t^2 + 76t - 280 = 0$$

$$-4(t^2 - 19t + 70) = 0$$

$$t^2 - 19t + 70 = 0$$

$$(t - 14)(t - 5) = 0$$

$t - 14 = 0$	$t - 5 = 0$
$t = 14$ min.	$t = 5$ min.

Section 1.4   (pages 31-33)

1.

$$x + 2 \le 0$$

$$x \le -2$$

$$(-\infty, -2]$$

3.

$$2x - 1 > 0$$

$$2x > 1$$

$$x > \frac{1}{2}$$

$$\left(\frac{1}{2}, \infty\right)$$

5.

$$7 + 2x > 3$$

$$2x > -4$$

$$x > -2$$

$$(-2, \infty)$$

7.

$$x - 4 \le 3x + 4$$

$$-2x \le 8$$

$$x \ge -4$$

$$[-4, \infty)$$

9.

$$2x + 1 < x + \frac{1}{2}$$

$$4x + 2 < 2x + 1$$

$$2x < -1$$

$$x < -\frac{1}{2}$$

$$\left(-\infty, -\frac{1}{2}\right)$$

11.

$$6x - 3(x + 1) > 0$$

$$6x - 3x - 3 > 0$$

$$3x - 3 > 0$$

$$3x > 3$$

$$x > 1$$

$$(1, \infty)$$

13.

$$2(x - 3) \le 3x$$

$$2x - 6 \le 3x$$

$$-x \le 6$$

$$x \ge -6$$

$$[-6, \infty)$$

15.

$$x^2 - 9 < 0$$

$$(x + 3)(x - 3) < 0$$

key points: $-3, 3$

Since $x^2 - 9 < 0$, the solution

is $-3 < x < 3$.

17. $x^2 + 7x + 12 \geq 0$

$(x+4)(x+3) \geq 0$

key points: $-4, -3$

$$\underset{\underset{-5 \quad -4 \quad -3 \quad -2 \quad -1}{}}{(-)(-) \quad (+)(-) \qquad (+)(+)}$$

Since $x^2 + 7x + 12 \geq 0$, the

solution is $x \leq -4$ or $x \geq -3$.

19. $x^2 + 2x < 8$

$x^2 + 2x - 8 < 0$

$(x+4)(x-2) < 0$

key points: $-4, 2$

$$\underset{\underset{-5 \quad -4 \quad -3 \quad -2 \quad -1 \quad 0 \quad 1 \quad 2 \quad 3}{}}{(-)(-) \qquad (+)(-) \qquad (+)(+)}$$

Since $x^2 + 2x - 8 < 0$, the solution

is $-4 < x < 2$.

21. $3x^2 - 10x - 8 \leq 0$

$(3x+2)(x-4) \leq 0$

key points: $-\frac{2}{3}, 4$

$$\underset{\underset{-2 \quad -1 \quad 0 \quad 1 \quad 2 \quad 3 \quad 4 \quad 5}{}}{(-)(-) \qquad (+)(-) \qquad (+)(+)}$$

Since $3x^2 - 10x - 8 \leq 0$, the

solution is $-\frac{2}{3} \leq x \leq 4$.

23. $4x^2 + 4x + 1 > 0$

$(2x+1)^2 > 0$

$(2x+1)^2 = 0$ for $x = -\frac{1}{2}$

and is positive for any other x .

Therefore, the solution is $x \neq -\frac{1}{2}$.

That is, $(-\infty, -\frac{1}{2}) \cup (-\frac{1}{2}, \infty)$.

25. $x^2 - 7 \geq 0$

$(x + \sqrt{7})(x - \sqrt{7}) \geq 0$

key points: $-\sqrt{7}, \sqrt{7}$

$$\underset{\underset{-4 \quad -3 \quad -2 \quad -1 \quad 0 \quad 1 \quad 2 \quad 3}{}}{(-)(-) \qquad (+)(-) \qquad (+)(+)}$$

Since $x^2 - 7 \geq 0$, the solution

is $x \leq -\sqrt{7}$ or $x \geq \sqrt{7}$.

27. $x^2 + x + 1 < 0$

To find key points:

$a = 1, b = 1$ and $c = 1$

$$x = \frac{-1 \pm \sqrt{1^2 - 4(1)(1)}}{2(1)}$$

$$= \frac{-1 \pm \sqrt{1 - 4}}{2}$$

$$= \frac{-1 \pm \sqrt{-3}}{2}$$

Therefore, there are no key points. This means the inequality is either always true or always false. Using $x = 2$ as a test point:

$(2)^2 + 2 + 1 < 0$

This is false. The inequality has no solution.

29. $2x^2 + 1 < 0$

To find key points:

$2x^2 + 1 = 0$

$2x^2 = -1$

$x^2 = -\frac{1}{2}$

Therefore there are no key points. This means the inequality is either always true or always false. Using $x = 1$ as a test point:

$2(1)^2 + 1 < 0$

This is false. The inequality has no solution.

31. $\frac{2}{x} > 0$

key point: 0

Since $\frac{2}{x} > 0$, the solution is $x > 0$.

33. $\frac{3x}{x-1} \leq 0$

key points: $3x = 0 \quad x - 1 = 0$

$\qquad\qquad x = 0 \quad x = 1$

Since $\frac{3x}{x-2} \leq 0$, the solution is $0 \leq x < 1$.

(Note that $x \neq 1$ since the denominator cannot be zero.)

35. $\frac{x+4}{x+2} \geq 0$

key points: $x + 4 = 0 \quad x + 2 = 0$

$\qquad\qquad x = -4 \qquad x = -2$

Since $\frac{x+4}{x+2} \geq 0$, the solution is

$x \leq -4$ or $x > -2$.

(Note that $x \neq -2$ since the denominator cannot be zero.)

37. $\frac{2x-1}{3-x} > 0$

key points:

$2x - 1 = 0 \qquad 3 - x = 0$

$x = \frac{1}{2} \qquad\qquad x = 3$

Since $\frac{2x-1}{3-x} > 0$, the solution is $\frac{1}{2} < x < 3$.

39. $\frac{x}{(x+1)(x-1)} < 0$

key points:

$x = 0 \qquad x + 1 = 0 \qquad x - 1 = 0$

$x = 0 \qquad x = -1 \qquad x = 1$

Since $\frac{x}{x^2-1} < 0$, the solution is $x < -1$ or $0 < x < 1$.

41. $\dfrac{x}{x-3} - 2 < 0$

$\dfrac{x}{x-3} - \dfrac{2(x-3)}{x-3} < 0$

$\dfrac{x - 2x + 6}{x-3} < 0$

$\dfrac{-x+6}{x-3} < 0$

key points:

$-x+6 = 0 \qquad x-3=0$

$x = 6 \qquad\qquad x = 3$

Since $\dfrac{x}{x-3} - 2 < 0$, the solution

is $x < 3$ or $x > 6$.

51. If $\dfrac{2}{\sqrt[3]{x}} > 0$ and $2 > 0$,

then $\sqrt[3]{x} > 0$

$x > 0$

55. $\dfrac{5-x}{\sqrt{x}} \geq 0$

(Since $\sqrt{x}$ cannot be negative

you do not need to check numbers

less than zero.)

key point: $5 - x = 0$

$x = 5$

no test points here

Since $\dfrac{5-x}{\sqrt{x}} \geq 0$, the solution

is $0 < x \leq 5$.

43. Since $\sqrt{x} > 0$, then $x > 0$.

45. If $\sqrt{x-2} \leq 0$ and $\sqrt{x-2}$

must also be positive or zero,

then $x = 2$.

47. If $\sqrt[3]{x+4} < 0$, then $x + 4 < 0$

$x < -4$

49. If $\sqrt{2x-1} > 0$, then $2x - 1 > 0$

$2x > 1$

$x > \dfrac{1}{2}$

53. If $\dfrac{-1}{\sqrt[3]{(x-1)^2}} < 0$

then $(x-2)^2 > 0$

$(x-1)^2 = 0$ for $x = 1$

and is positive for any other x.

Therefore, the solution is $x \neq 1$.

That is, $(-\infty, 1) \cup (1, \infty)$.

57. $\dfrac{x+3}{\sqrt{x-3}} < 0$

Since $\sqrt{x-3}$ must be positive,

then x must be greater than 3.

However, if $\dfrac{x+3}{\sqrt{x-3}} < 0$,

then $x + 3 < 0$

and $x < -3$.

Since there is no number greater

than 3 and less than -3, the

inequality has no solution.

59. $\dfrac{1}{\sqrt[4]{2x}} > 0$, so $\sqrt[4]{2x} > 0$, $2x > 0$, $x > 0$

61. $(x+1)(x-2)(x+3) > 0$

key points: $x+1=0 \qquad x-2=0 \qquad x+3=0$

$\qquad\qquad\quad x=-1 \qquad x=2 \qquad x=-3$

Since $(x+1)(x-2)(x+3) > 0$, the solution is $-3 < x < -1$ or $x > 2$.

63. $x(x-5)^2 \le 0$

key points: $x=0 \quad (x-5)^2 = 0$

$\qquad\qquad\qquad\qquad x=5$

Since $x(x-5)^2 \le 0$, and

$x(x-5)^2 = 0$

when $x=5$, the solution is

$x \le 0$ or $x = 5$.

65. $x^3 - x^2 - 6x \ge 0$

$x(x^2 - x - 6) \ge 0$

$x(x-3)(x+2) \ge 0$

key points:

$x = 0 \qquad x-3=0 \qquad x+2+0$

$\qquad\qquad\quad x=3 \qquad\quad x=-2$

Since $x^3 - x^2 - 6x \ge 0$, the solution

is $-2 \le x \le 0$ or $x \ge 3$.

67a. $\qquad -5x^2 + 55x - 50 = 0$

$\qquad -5(x^2 - 11x + 10) = 0$

$\qquad\qquad x^2 - 11x + 10 = 0$

$\qquad\qquad (x-1)(x-10) = 0$

$x - 1 = 0 \qquad\quad x - 10 = 0$

$x = 1 \qquad\qquad x = 10$

Since $x < 10$, the profit is 0 at

$x = 1$ (100 boxes).

67b. $\qquad -5x^2 + 55x - 50 > 0$

$\qquad -5(x^2 - 11x + 10) > 0$

$\qquad -5(x-1)(x-10) > 0$

key points: $x-1=0 \qquad x-10=0$

$\qquad\qquad\qquad x=1 \qquad\qquad x=10$

No test points for $x < 0$ or $x > 10$.

Since $-5x^2 + 55x - 50 > 0$, the

solution is $1 < x < 10$ (more than

100 boxes but less than 1000 boxes).

Section 1.5 (pages 46-48)

1. $|0-(-4)| = |4|$
    $= 4$

3. $|2-(-5)| = |7|$
    $= 7$

5. $|-3-(-8)| = |5|$
    $= 5$

7.

9.

11. $\sqrt{(-2-2)^2 + (6-3)^2} = \sqrt{(-4)^2 + (3)^2}$
    $= \sqrt{16+9}$
    $= \sqrt{25}$
    $= 5$

13. $\sqrt{[-5-(-2)]^2 + (-2-1)^2} = \sqrt{(-3)^2 + (-3)^2}$
    $= \sqrt{9+9}$
    $= \sqrt{18}$
    $= 3\sqrt{2}$

15. $\sqrt{(-1-0)^2 + (3-1)^2} = \sqrt{(-1)^2 + (2)^2}$
    $= \sqrt{1+4}$
    $= \sqrt{5}$

17. $\sqrt{[1.2-(-3.2)]^2+[-1-(-1)]^2} = \sqrt{(4.4)^2+(0)^2}$
$$= \sqrt{(4.4)^2}$$
$$= 4.4$$

19. $m = \dfrac{8-2}{2-4}$

$\quad = \dfrac{6}{-2}$

$\quad = -3$

21. $m = \dfrac{-6-(-1)}{2-(-3)}$

$\quad = \dfrac{-5}{5}$

$\quad = -1$

23. $m = \dfrac{0-3}{-2-0}$

$\quad = \dfrac{-3}{-2}$

$\quad = \dfrac{3}{2}$

25. $m = \dfrac{-1-(-2)}{1-5}$

$\quad = -\dfrac{1}{4}$

27. $m = \dfrac{3-3}{-5-0}$

$\quad = \dfrac{0}{-5}$

$\quad = 0$

29. $m = \dfrac{7-(-1)}{-2-(-2)}$

$\quad = \dfrac{8}{0}$

$\quad$ undefined

31. $y-y_1 = m(x-x_1)$

$y-4 = 6(x-3)$

$y-4 = 6x-18$

$y = 6x-14$

33. $y-y_1 = m(x-x_1)$

$y-4 = -3[x-(-1)]$

$y-4 = -3x-3$

$y = -3x+1$

35. $y-y_1 = m(x-x_1)$

$y-4 = \frac{1}{2}(x-5)$

$y-4 = \frac{1}{2}x-\frac{5}{2}$

$y = \frac{1}{2}x+\frac{3}{2}$

37. $y-y_1 = m(x-x_1)$

$y-\frac{5}{4} = -\frac{2}{3}(x-\frac{1}{2})$

$y-\frac{5}{4} = -\frac{2}{3}x+\frac{1}{3}$

$y = -\frac{2}{3}x+\frac{19}{12}$

39. $y-y_1 = m(x-x_1)$

$y-5 = 0(x-3)$

$y-5 = 0$

$y = 5$

41. If m is undefined, then the line is vertical. The equation is $x = -2$.

43.
$$m = \frac{6-3}{-2-1}$$
$$= \frac{3}{-3}$$
$$= -1$$

To find the equation of the line, use either given point in the point-slope form.

For $(1,3)$:
$$y - 3 = -1(x-1)$$
$$y - 3 = -x + 1$$
$$x + y - 4 = 0$$

45.
$$m = \frac{5-5}{0-5}$$
$$= \frac{0}{-5}$$
$$= 0$$

If $m = 0$, then the line is horizontal. The equation is $y = 5$ or $y - 5 = 0$.

47.
$$m = \frac{8-3}{-4-(-2)}$$
$$= -\frac{5}{2}$$

Use either given point in the point-slope form.

For $(-2,3)$:
$$y - 3 = -\frac{5}{2}[x - (-2)]$$
$$2(y-3) = -5(x+2)$$
$$2y - 6 = -5x - 10$$
$$5x + 2y + 4 = 0$$

49. $y = -4x + 6$, $m = -4$, $b = 6$

51.
$$x - 2y + 6 = 0$$
$$-2y = -x - 6$$
$$y = \tfrac{1}{2}x + 3$$
$$m = \tfrac{1}{2}, \ b = 3$$

53.
$$5x + 4y = 0$$
$$4y = -5x$$
$$y = \frac{-5}{4}x$$
$$m = -\tfrac{5}{4}, \ b = 0$$

55. $y = 0x + 6$, $m = 0$, $b = 6$

57.

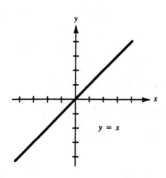

$y = x$

63.

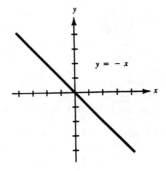

$y = -x$

69.

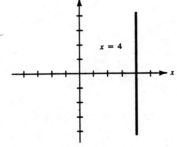

$x = 4$

71. Since parallel lines have the same slope, $m = 4$.

$$y - (-3) = 4[x - (-2)]$$
$$y + 3 = 4x + 8$$
$$0 = 4x - y + 5$$

73.
$$x + 2y = 0$$
$$2y = -x$$
$$y = -\tfrac{1}{2}x$$
$$m = -\tfrac{1}{2}$$
$$y - 5 = -\tfrac{1}{2}(x - 0)$$
$$2(y - 5) = -1(x - 0)$$
$$2y - 10 = -x$$
$$x + 2y - 10 = 0$$

75. If $x + 3 = 0$, then the line is vertical:

$$x = 5$$
$$\text{or } x - 5 = 0$$

77. The slope of the line $y = 5x - 2$ is $m = 5$.

A line perpendicular to it has slope $m = -\tfrac{1}{5}$.

$$y - 1 = -\tfrac{1}{5}(x - 1)$$
$$5(y - 1) = -1(x - 1)$$
$$5y - 5 = -x + 1$$
$$x + 5y - 6 = 0$$

79. $x + 3y + 4 = 0$

$$3y = -x + 4$$

$$y = -\frac{1}{3}x + \frac{4}{3}$$

The slope of this line is $m = -\frac{1}{3}$.
A line perpendicular to it has
slope $m = 3$.

$$y - (-1) = 3(x - 0)$$

$$y + 1 = 3x$$

$$0 = 3x - y - 1$$

81. Substitute $y = 2x - 1$ into the
equation $x - y + 4 = 0$.

$$x - (2x - 1) + 4 = 0$$

$$x - 2x + 1 + 4 = 0$$

$$-x + 5 = 0$$

$$-x = -5$$

$$x = 5$$

To find y: $\quad y = 2x - 1$

$$y = 2(5) - 1$$

$$y = 9$$

Answer: $(5, 9)$

83. Substitute $y = 2x - 5$ into the
equation $y = 3x - 3$.

$$2x - 5 = 3x - 3$$

$$-2 = x$$

To find y: $\quad y = 2x - 5$

$$y = 2(-2) - 5$$

$$y = -9$$

Answer: $(-2, -9)$

85. Solve the equation $2x - y + 5 = 0$
for y: $\quad y + 2x + 5$

Substitute this into the equation
$x + 2y - 5 = 0$.

$$x + 2(2x + 5) - 5 = 0$$

$$x + 4x + 10 - 5 = 0$$

$$5x + 5 = 0$$

$$5x = -5$$

$$x = -1$$

To find y: $\quad y = 2x + 5$

$$y = 2(-1) + 5$$

$$y = 3$$

Answer: $(-1, 3)$

87. $\qquad p = 2000 - 2x$

$$20 = 2000 - 2x$$

$$-1980 = -2x$$

$$990 = x$$

89. $\qquad p = 740 - 0.2x^2$

$$20 = 740 - 0.2x^2$$

$$0.2x^2 = 720$$

$$x^2 = 3600$$

Since x cannot be negative, $x = 60$.

91.

$$p = 4x + 1200$$
$$1500 = 4x + 1200$$
$$300 = 4x$$
$$75 = x$$

93.

$$p = 0.02x^2 + 350$$
$$400 = 0.02x^2 + 350$$
$$50 = 0.02x^2$$
$$2500 = x^2$$

Since x cannot be negative, $x = 50$.

95. Substitute $p = 0.2x + 40$ into the equation $p = 50 - 0.3x$

$$0.2x + 40 = 50 - 0.3x$$
$$0.5x = 10$$
$$x = 20$$
$$p = 0.2x + 40$$
$$p = 0.2(20) + 40$$
$$p = 44$$

Answer: 20 units, \$44

97. Substitute $p = 0.1x^2 + 200$ into the equation $p = 400 - x$

$$0.1x^2 + 200 = 400 - x$$
$$0.1x^2 + x - 200 = 0$$
$$x^2 + 10x - 2000 = 0$$
$$(x - 40)(x + 50) = 0$$

$x - 40 = 0$	$x + 50 = 0$
$x = 40$	$x = -50$

Since x cannot be negative, $x = 40$.

$$p = 400 - x$$
$$p = 400 - 40$$
$$p = 360$$

Answer: 40 units, \$360

99. When $x = 500$, then $p = 7$.

When $x = 500 + 200 = 700$, then $p = 7 - 1 = 6$.

Using $(500, 7)$ and $(700, 6)$

$$m = \frac{6 - 7}{700 - 500}$$
$$= \frac{-1}{200} = -0.05$$
$$p - p_1 = m(x - x_1)$$
$$p - 7 = -0.005(x - 500)$$
$$p - 7 = -0.005x + 2.5$$
$$p = -0.005x + 9.5$$

Chapter 1 Review (pages 51-52)

1. $3x^3 + 6x^2 - 5x^2 + 20x$

$= 3x^3 + x^2 + 20x$

2. $\dfrac{2x(x+2)}{(x-1)(x+2)} + \dfrac{3(x-1)}{(x+2)(x-1)}$

$= \dfrac{2x^2 + 4x + 3x - 3}{(x-1)(x+2)}$

$= \dfrac{2x^2 + 7x - 3}{(x-1)(x+2)}$

3. $x^{-1/2} \cdot x + x^{-1/2} \cdot x^{1/2}$

$= x^{1/2} + x^0$

$= x^{1/2} + 1$

4. $2(x^2 + x - 20)$

$= 2(x+5)(x-4)$

5. $2(x+4)^4[5x + (x+4)]$

$= 2(x+4)^4(6x+4)$

$= 4(x+4)^4(3x+2)$

6. $(x^4 + 1)(x^4 - 1)$

$(x^4 + 1)(x^2 + 1)(x^2 - 1)$

$= (x^4 + 1)(x^2 + 1)(x+1)(x-1)$

7. $(2y)^3 - (3)^3$

$= (2y - 3)(4y^2 + 6y + 9)$

8. $x^{-3}(1 - 4x)$

$= \dfrac{1 - 4x}{x^3}$

9. $4x + 6x - 12 = 0$

$10x = 12$

$x = \dfrac{6}{5}$

10. $x(x - 8) = 0$

$x = 0$ $\quad\bigg|\quad$ $x - 8 = 0$

$\qquad\qquad\quad x = 8$

11. $\sqrt{(x-3)^2} = \pm\sqrt{16}$

$x - 3 = \pm 4$

$x - 3 = 4$ $\quad\bigg|\quad$ $x - 3 = -4$

$x = 7$ $\qquad\quad\bigg|\quad$ $x = -1$

12. $x^2 - 5x - 1 = 0$

$a = 1, b = -5$ and $c = -1$

$x = \dfrac{-(-5) \pm \sqrt{(-5)^2 - 4(1)(-1)}}{2(1)}$

$x = \dfrac{5 \pm \sqrt{25 + 4}}{2}$

$x = \dfrac{5 \pm \sqrt{29}}{2}$

13. $(\sqrt[3]{3x-1})^3 = (1)^3$

    $3x - 1 = 1$

    $3x = 2$

    $x = \dfrac{2}{3}$

14. $(2x-3)\left(\dfrac{2x+1}{2x-3}\right) = (2x-3)(2)$

    $2x + 1 = 4x - 6$

    $-2x = -7$

    $x = \dfrac{7}{2}$

15. $\dfrac{x^2 - x}{\sqrt{x+1}} = 0$

    $x^2 - x = 0$

    $x(x-1) = 0$

    $x = 0 \quad\Big|\quad x - 1 = 0$

                       $x = 1$

16. $1 - 3x < -2x + 7$

    $-x < 6$

    $x > -6$

    $(-6, \infty)$

17. $x^2 - 4 \geq 0$

    $(x+2)(x-2) \geq 0$

    key points:

    $x + 2 = 0 \qquad x - 2 = 0$

    $x = -2 \qquad\;\; x = 2$

    $(-)(-) \qquad (+)(-) \qquad (+)(+)$

    $-4\;\;-3\;\;-2\;\;-1\;\;0\;\;1\;\;2\;\;3\;\;4$

    Since $x^2 - 4 \geq 0$, the solution

    is $x \leq -2$ or $x \geq 2$.

18. $x^2 - 2x \leq 0$

    $x(x-2) \leq 0$

    key points:

    $x = 0 \qquad x - 2 = 0$

                    $x = 2$

    $(-)(-) \quad (+)(-) \quad (+)(+)$

    $-2\;\;-1\;\;0\;\;1\;\;2\;\;3\;\;4$

    Since $x^2 - 2x \leq 0$, the solution

    is $0 \leq x \leq 2$.

19. $\dfrac{3x+1}{x} < 0$

    key points: $3x + 1 = 0 \qquad x = 0$

                        $x = -\dfrac{1}{3}$

    $\dfrac{(-)}{(-)}\;\;\dfrac{(+)}{(-)}\;\;\dfrac{(+)}{(+)}$

    $-2\;\;\;\;-1\;\;\;\;0\;\;\;\;1$

    Since $\dfrac{3x+1}{x} < 0$, the solution

    is $-\dfrac{1}{3} < x < 0$.

20. $\sqrt{4-x} \geq 0$

    $4 - x \geq 0$

    $-x \geq -4$

    $x \leq 4$

21. $\sqrt{[4-(-2)]^2 + (13-5)^2}$

    $= \sqrt{6^2 + 8^2}$

    $= \sqrt{36 + 64}$

    $= \sqrt{100}$

    $= 10$

22. $m = \dfrac{8-6}{-1-0}$

$= \dfrac{2}{-1}$

$= -2$

Use either given point in the point-slope form.

For $(0,6)$:

$y - 6 = -2(x - 0)$

$y - 6 = -2x$

$2x + y - 6 = 0$

23. $2x + 4y + 7 = 0$

$4y = -2x - 7$

$y = -\dfrac{1}{2}x - \dfrac{7}{4}$

$m = -\dfrac{1}{2}, \; b = -\dfrac{7}{4}$

24. $p = 250 - 0.2x$

$75 = 250 - 0.2x$

$0.2x = 175$

$x = 875$

25. $0 = -125 + 30x - x^2$

$x^2 - 30x + 125 = 0$

$(x - 5)(x - 25) = 0$

$x - 5 = 0 \qquad\qquad x - 25 = 0$

$x = 5 \qquad\qquad\quad x = 25$

Since $x \le 20$, the answer is $x = 5$

(500 cards).

Chapter 1 Test   (pages 52-54)

1. $\dfrac{2}{\sqrt{2x}} \cdot \dfrac{\sqrt{2x}}{\sqrt{2x}}$

$= \dfrac{2\sqrt{2x}}{\sqrt{(2x)^2}}$

$= \dfrac{2\sqrt{2x}}{2x}$

$= \dfrac{\sqrt{2x}}{x}$ (b)

2. $x^4 - 4x^2$

$= x^2(x^2 - 4)$

$= x^2(x + 2)(x - 2)$ (b)

3. $\left(\sqrt[3]{x^2 + 1}\right)^3 = (2)^3$

$x^2 + 1 = 8$

$x^2 = 7$

$x = \pm\sqrt{7}$ (d)

4. $x^3 + 7x^2 - 5x - 35$ (b)

5. $a = 1$, $b = 4$ and $c = -2$

$$x = \frac{-4 \pm \sqrt{(4)^2 - 4(1)(-2)}}{2(1)}$$

$$x = \frac{-4 \pm \sqrt{16 + 8}}{2}$$

$$x = \frac{-4 \pm \sqrt{24}}{2}$$

$$x = \frac{-4 \pm 2\sqrt{6}}{2}$$

$$x = \frac{2(-2 \pm \sqrt{6})}{2}$$

$$x = -2 \pm \sqrt{6} \qquad \text{(d)}$$

6. $\dfrac{x}{x+3} > 0$

key points: $x = 0 \qquad x + 3 = 0$

$$x = -3$$

Since $\dfrac{x}{x+3} > 0$, the solution

is $x < -3$ or $x > 0$.   (b)

7. $\sqrt{(5-3)^2 + (-1-2)^2}$

$= \sqrt{2^2 + (-3)^2}$

$= \sqrt{4 + 9}$

$= \sqrt{13} \qquad \text{(d)}$

8. $y - y_1 = m(x - x_1)$

$y - (-4) = \frac{1}{2}(x - 5)$

$2(y + 4) = 1(x - 5)$

$2y + 8 = x - 5$

$0 = x - 2y - 13 \qquad \text{(b)}$

9. $y = -2x + 1$

$m = -2$

$b = 1 \qquad \text{(c)}$

10. A line perpendicular to the x-axis is
a vertical line:

$x = -6$

$x + 6 = 0 \qquad \text{(a)}$

11. $3 - 2x = 0$

$-2x = -3$

$x = \frac{3}{2}$

12. $-2(16)^{-3/4} = \dfrac{-2}{16^{3/4}}$

$= \dfrac{-2}{\left(\sqrt[4]{16}\right)^3} = \dfrac{-2}{2^3}$

$= \dfrac{-2}{8} = -\dfrac{1}{4}$

13. $x^2 - 4x \geq 0$

$x(x-4) \geq 0$

key points: $x = 0$ $x - 4 = 0$

                          $x = 4$

Since $x^2 - 4x \geq 0$, the solution

is $x \leq 0$ or $x \geq 4$.

14. $2x(x-1)^{-3}[(x-1)+2x]$

$= 2x(x-1)^{-3}(3x-1)$

$= \dfrac{2x(3x-1)}{(x-1)^3}$

15. $m = \dfrac{-3-(-4)}{0-2}$

$ = -\dfrac{1}{2}$

Use either given point in the

point-slope form.

For $(0, -3)$:

$y - (-3) = -\dfrac{1}{2}(x - 0)$

$2(y+3) = -1(x-0)$

$2y + 6 = -x$

$x + 2y + 6 = 0$

16. $x^2 - 2x = 0$

$x(x-2) = 0$

$x = 0$ | $x - 2 = 0$

                          $x = 2$

17. $3.50 = 12.5 - 0.01x$

        $0.01x = 9$

           $x = 900$

18. $0.1x^2 - 250 < 0$

           $x^2 - 2500 < 0$

$(x+50)(x-50) < 0$

key point: $x - 50 = 0$

           $x = 50$ $(x \geq 0)$

Since $0.1x^2 - 250 < 0$, the company will

lose money for $0 < x < 50$. It will make

money for $x > 50$.

CHAPTER 2: FUNCTIONS, LIMITS, AND THE DERIVATIVE

Section 2.1 (pages 69-72)

1. Yes, each first component is different.

3. Yes, each first component is different.

5. Yes, there is only one y value for each x value.

7. Yes, there is only one y value for each x value.

9. Yes, there is only one y value for each x value.

11. domain: $\{-3, -2, -1\}$
    range: $\{1, 2, 3\}$

13. domain: all real numbers
    range: all real numbers

15. Since $f(x) = \sqrt{x}$, x cannot be negative.
    domain: $x \geq 0$
    range: $y \geq 0$

17. domain: $x \geq 0$
    range: $y \geq 0$

19a. $f(2) = 5$

19b. $f(-3) = 5$

21a. $f(-4) = 7 - 4(-4) = 23$

23a. $g(-1) = 3(-1)^2 + (-1) - 4 = -2$

21b. $f(1) = 7 - 4(1) = 3$

23b. $g(0) = 3(0)^2 + 0 - 4 = -4$

21c. $f(0) = 7 - 4(0) = 7$

23c. $g(-x) = 3(-x)^2 + (-x) - 4$
    $= 3x^2 - x - 4$

21d. $f(n) = 7 - 4h$

23d. $g(x + h) = 3(x + h)^2 + (x + h) - 4$
    $= 3(x^2 + 2xh + h^2) + (x + h) - 4$
    $= 3x^2 + 6xh + 3h^2 + x + h - 4$

25a. $f(x + h) = 4(x + h) + 6$
    $= 4x + 4h + 6$

25b. $f(x + h) - f(x) = (4x + 4h + 6) - (4x + 6)$
    $= 4h$

25c. $\dfrac{f(x + h) - f(x)}{h} = \dfrac{4h}{h} = 4$

27a. $f(x+h) = (x+h)^2 + (x+h)$
$$= x^2 + 2xh + h^2 + x + h$$

27b. $f(x+h) - f(x) = (x^2 + 2xh + h^2 + x + h) - (x^2 + x)$
$$= 2xh + h^2 + h$$

27c. $\dfrac{f(x+h) - f(x)}{h} = \dfrac{2xh + h^2 + h}{h}$
$$= \dfrac{h(2x + h + 1)}{h}$$
$$= 2x + h + 1$$

29a. $f(x+h) = \dfrac{1}{x+h}$

29b. $f(x+h) - f(x) = \dfrac{1}{x+h} - \dfrac{1}{x}$
$$= \dfrac{x}{(x+h)x} - \dfrac{x+h}{x(x+h)}$$
$$= \dfrac{-h}{x(x+h)}$$

29c. $\dfrac{f(x+h) - f(x)}{h} = \dfrac{1}{h} \cdot \dfrac{-h}{x(x+h)}$
$$= \dfrac{-1}{x(x+h)}$$

31.

x	y
-1	4
0	4
1	4

33.

x	y
-2	-4
0	0
2	4

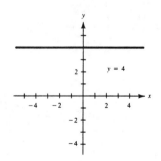

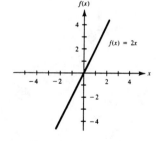

35.

x	y
-1	8
0	5
1	2

$y = 5 - 3x$

37.

x	y
-3	-9
-2	-4
-1	-1
0	0
1	-1
2	-4
3	-9

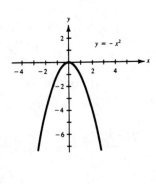

$y = -x^2$

39.

x	y
-6	9
-5	4
-4	1
-3	0
-2	1
-1	4
0	9

$f(x) = (x + 3)^2$

41.

x	y
0	0
1	1
4	2
9	3

$y = \sqrt{x}$

43.

x	y
-3	$\frac{1}{3}$
-2	$\frac{1}{2}$
-1	1
$-\frac{1}{2}$	2
$\frac{1}{2}$	-2
1	-1
2	$-\frac{1}{2}$
3	$-\frac{1}{3}$

$y = \dfrac{-1}{x}$

45.

x	y
-2	-8
-1	-1
0	0
1	1
2	8

$y = x^3$

47.

x	y
0	0
0.5	0
0.9	0
1	-1
1.5	-1
1.9	-1
2	-2

$y = -[x]$

49.

x	y
-2	4
-1	3
0	2
1	3
2	4

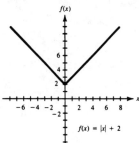

$f(x) = |x| + 2$

51.

x	y
-2	2
-1	1
0	0
1	-1
2	2
3	3

$$f(x) = \begin{cases} -x, & x \le 1 \\ x, & x > 1 \end{cases}$$

53.

x	y
-3	-8
-2	-3
-1	0
0	1
1	2
2	3

$$y = \begin{cases} 1 - x^2, & x < 0 \\ x + 1, & x \ge 0 \end{cases}$$

55.
$$\begin{aligned} f(x) + g(x) &= (2x^2 - x) + (3x + 4) \\ &= 2x^2 + 2x + 4 \end{aligned}$$

57.
$$\begin{aligned} f(x) \cdot g(x) &= (2x^2 - x)(3x + 4) \\ &= 6x^3 + 8x^2 - 3x^2 - 4x \\ &= 6x^3 + 5x^2 - 4x \end{aligned}$$

59.
$$f(1) = 2(1)^2 - 1 = 1$$
$$g(0) = 3(0) + 4 = 4$$
$$\frac{f(1)}{g(0)} = \frac{1}{4}$$

61.
$$g(-2) = 3(-2) + 4 = -2$$
$$f[g(-2)] = f(-2)$$
$$= 2(-2)^2 - (-2)$$
$$= 10$$

63.
$$f[g(x)] = f(3x)$$
$$= 2 - 3x$$
$$g[f(x)] = g(2 - x)$$
$$= 3(2 - x)$$
$$= 6 - 3x$$

65.
$$f[g(x)] = f(x + 1)$$
$$= (x + 1)^2 + (x + 1)$$
$$= x^2 + 2x + 1 + x + 1$$
$$= x^2 + 3x + 2$$
$$g[f(x)] = g(x^2 + x)$$
$$= (x^2 + x) + 1$$
$$= x^2 + x + 1$$

67.
$$f[g(x)] = f(4 - x)$$
$$= 4 - (4 - x)$$
$$= x$$
$$g[f(x)] = g(4 - x)$$
$$= 4 - (4 - x)$$
$$= x$$

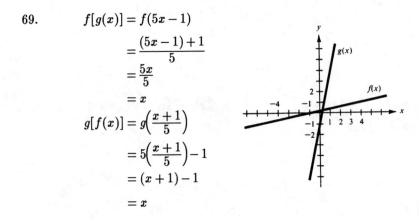

69.
$$f[g(x)] = f(5x - 1)$$
$$= \frac{(5x - 1) + 1}{5}$$
$$= \frac{5x}{5}$$
$$= x$$
$$g[f(x)] = g\left(\frac{x + 1}{5}\right)$$
$$= 5\left(\frac{x + 1}{5}\right) - 1$$
$$= (x + 1) - 1$$
$$= x$$

71.
$$f[g(x)] = f(\sqrt[3]{x})$$
$$= (\sqrt[3]{x})^3$$
$$= x$$
$$g[f(x)] = g(x^3)$$
$$= \sqrt[3]{x^3}$$
$$= x$$

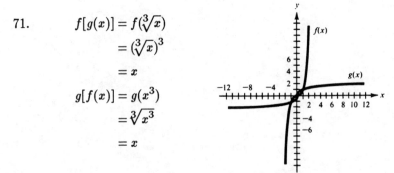

73.

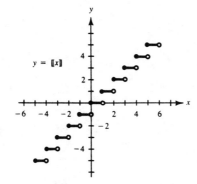

75.
$$A = lw$$
$$108 = lw$$
$$\frac{108}{w} = l$$
$$P = 2w + 2l$$
$$P(w) = 2w + 2\left(\frac{108}{w}\right)$$
$$P(w) = 2w + \frac{216}{w}$$

77. No. For example, when $x = 2$, $y = 4$ and when $x = -2$, $y = 4$.

79. Yes. For every y, there is exactly one x.

81.
$$\begin{aligned} C(100) &= 800 + 0.1(100)^2 \\ &= \$1800 \end{aligned}$$

83.
$$\begin{aligned} C(200) &= 1000 + 0.1(200)^3 \\ &= \$801,000 \end{aligned}$$

85.
$$\begin{aligned} R(100) &= 70(100) - 0.5(100)^2 \\ &= \$2000 \end{aligned}$$

87.
$$\begin{aligned} R(100) &= 3000(100) - 0.2(100)^3 \\ &= \$100,000 \end{aligned}$$

89.
$$\begin{aligned} P(x) &= (40x - x^2) - (100 + 12x) \\ &= -x^2 + 28x - 100 \end{aligned}$$

91.
$$\begin{aligned} P(x) &= (1.75x^2 - 0.5x^3) - (100 + 0.25x^2) \\ &= -0.5x^3 + 1.5x^2 - 100 \end{aligned}$$

93.
$$\begin{aligned} R(500) &= 13(500) - 0.01(500)^2 \\ &= \$4000 \end{aligned}$$

95. height $= y$, length $=$ width $= x$
$$V = lwh$$
$$64 = x \cdot x \cdot y$$
$$\frac{64}{x^2} = y$$

The surface area of each of the four sides of the box is xy. The surface area of the top and the bottom of the box is x^2.

The surface area of the box is:
$$S = 2x^2 + 4xy, \quad \text{so} \quad S(x) = 2x^2 + 4x\left(\frac{64}{x^2}\right) = 2x^2 + \frac{256}{x}$$

97. $B(3) = 500 + 4(3)^2$ $B(0) = 500 + 4(0)^3$
           $= 608$ $= 500$

To find the time for the bacteria to double:

$$1000 = 500 + 4t^3, \ 500 = 4t^3, \ 125 = t^3, \ 5 = t$$

Section 2.2   (pages 83-88)

1. $\lim_{x \to 3} (2x - 6) = 0$

3. $\lim_{x \to -2} \dfrac{1}{x + 3} = 1$

5. $\lim_{x \to 3} 5 = 5$ 7. $\lim_{x \to -1} (2x + 1) = 2(-1) + 1 = -1$

9. $\lim_{x \to 0} (x^2 + 2x) = 0^2 + 2(0) = 0$ 11. $\lim_{x \to 2} \dfrac{x + 1}{x - 4} = \dfrac{2 + 1}{2 - 4} = \dfrac{3}{-2}$

13. $\lim_{x \to -2} \dfrac{5x}{x + 2} = \dfrac{5(-2)}{-2 + 2}$ 15. $\lim_{x \to 7} \sqrt{2x - 5} = \sqrt{2(7) - 5}$

          $= \dfrac{-10}{0}$ $= \sqrt{9} = 3$

       limit does not exist

17. $\lim_{x \to 9} \dfrac{1}{3 - \sqrt{x}} = \dfrac{1}{3 - \sqrt{9}} = \dfrac{1}{0}$ 19. $\lim_{x \to -4} (x^2 + \tfrac{1}{x}) = (-4)^2 + \dfrac{1}{-4}$

       limit does not exist $= 16 - \tfrac{1}{4} = 15\tfrac{3}{4}$

21. $\lim_{x \to 1/2} \dfrac{x - 1}{2x + 3} = \dfrac{\frac{1}{2} - 1}{2(\frac{1}{2}) + 3} = \dfrac{\frac{-1}{2}}{1 + 3}$ 23. $\lim_{x \to 1^+} (x + 3) = 1 + 3 = 4$

          $= \dfrac{-1}{2} \cdot \dfrac{1}{4} = -\dfrac{1}{8}$

25. $\lim_{x \to 2} (3x^2 + x - 1) = 3(2)^2 + 2 - 1$ 27. $\lim_{x \to -4^+} \dfrac{x + 4}{x + 1} = \dfrac{-4 + 4}{-4 + 1} = \dfrac{0}{-3} = 0$

          $= 13$

29. $\lim_{x \to 0^+} (5 - \sqrt{x}) = 5 - \sqrt{0} = 5$

31.

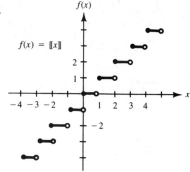

$f(x) = [[x]]$

33.

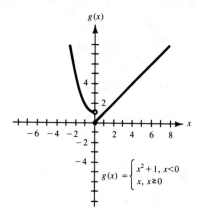

$$g(x) = \begin{cases} x^2+1, & x<0 \\ x, & x\ge 0 \end{cases}$$

31a. $\displaystyle\lim_{x\to 0^+} f(x) = 0$

33a. $\displaystyle\lim_{x\to -2} g(x) = (-2)^2 + 1 = 5$

31b. $\displaystyle\lim_{x\to 0^-} f(x) = -1$

33b. $\displaystyle\lim_{x\to 3} g(x) = 3$

31c. $\displaystyle\lim_{x\to 0} f(x)$ does not exist

33c. $\displaystyle\lim_{x\to 0^-} g(x) = 0^2 + 1 = 1$

since $\displaystyle\lim_{x\to 0^+} f(x) \ne \lim_{x\to 0^-} f(x)$

33d. $\displaystyle\lim_{x\to 0^+} g(x) = 0$

31d. $\displaystyle\lim_{x\to 1} f(x)$ does not exist

since $\displaystyle\lim_{x\to 1^-} f(x) = 0$ and

$\displaystyle\lim_{x\to 1^+} f(x) = 1$

35.

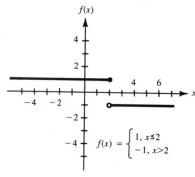

$$f(x) = \begin{cases} 1, & x\le 2 \\ -1, & x>2 \end{cases}$$

37. $\displaystyle\lim_{x\to 3} \frac{(x+3)(x-3)}{x-3}$

$= \displaystyle\lim_{x\to 3} (x+3) = 3 + 3 = 6$

39. $\displaystyle\lim_{x\to -2} \frac{x+2}{x(x+2)} = \lim_{x\to -2} \frac{1}{x} = \frac{-1}{2}$

41. $\displaystyle\lim_{x\to 0} \frac{x^2(x^2-1)}{x^2}$

$= \displaystyle\lim_{x\to 0} (x^2-1) = 0^2 - 1 = -1$

35a. $\displaystyle\lim_{x\to 2^-} f(x) = 1$

35b. $\displaystyle\lim_{x\to 2^+} f(x) = -1$

35c. $\displaystyle\lim_{x\to 2} f(x)$ does not exist since $\displaystyle\lim_{x\to 2^-} f(x) \ne \lim_{x\to 2^+} f(x)$

35d. $\displaystyle\lim_{x\to 0} f(x) = 1$

43. $\displaystyle\lim_{x\to 9}\frac{\sqrt{x}-3}{x-9}\cdot\frac{\sqrt{x}+3}{\sqrt{x}+3}$

$\displaystyle=\lim_{x\to 9}\frac{x-9}{(x-9)(\sqrt{x}+3)}$

$\displaystyle=\lim_{x\to 9}\frac{1}{\sqrt{x}+3}=\frac{1}{\sqrt{9}+3}=\frac{1}{6}$

45. $\displaystyle\lim_{x\to -5^+}\frac{x+5}{\sqrt{x+5}}\cdot\frac{\sqrt{x+5}}{\sqrt{x+5}}$

$\displaystyle=\lim_{x\to -5^+}\frac{(x+5)\sqrt{x+5}}{x+5}$

$\displaystyle=\lim_{x\to -5^+}\sqrt{x+5}=\sqrt{-5+5}=0$

47. $\displaystyle\lim_{x\to 0}\frac{x^2+x}{x}=\lim_{x\to 0}\frac{x(x+1)}{x}$

$\displaystyle=\lim_{x\to 0}(x+1)=0+1=1$

49. $\displaystyle\lim_{x\to 1/2}\frac{(x+1)(2x-1)}{2x-1}$

$\displaystyle=\lim_{x\to 1/2}(x+1)=\frac{1}{2}+1=\frac{3}{2}$

51a.

51b. $\displaystyle\lim_{x\to 0}f(x)=-3$

51c. They are the same.

53a. $\displaystyle\lim_{t\to 10^-}P(t)=300$

53b. $\displaystyle\lim_{t\to 10^+}P(t)=200$

55a. $\displaystyle\lim_{x\to 4000}T(x)=130.90+.12(4000-3670)=\$170.50$

55b. $\displaystyle\lim_{x\to 50,000}T(x)=11,134.20+.42(50,000-44,780)=\$13,326.60$

55c. $\displaystyle\lim_{x\to 31,080^-}T(x)=4441.00+.30(31,080-25,360)=\$6157.00$

55d. $\displaystyle\lim_{x\to 31,080^+}T(x)=6157.00+.34(31,080-31,080)=\$6157.00$

57. $\displaystyle\lim_{t\to 35}N(t)=1600-(35-75)^2$

$=0$

Section 2.3 (pages 93-96)

1. continuous

3. 0

5. -2 and 2

7. 1

9. discontinuous at every integer

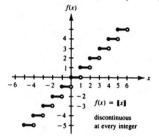

$f(x) = \llbracket x \rrbracket$

discontinuous at every integer

11. discontinuous at $x = 0$

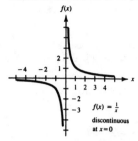

$f(x) = \frac{1}{x}$

discontinuous at $x = 0$

13. discontinuous at $x = -1$

15. continuous

17. discontinuous at $x = 1$

19. $f(4) = 1 - 2(4) = -7$

$$\lim_{x \to 4} f(x) = \lim_{x \to 4} (1 - 2x) = -7$$

$\lim_{x \to 4} f(x) = f(4)$ so the function is continuous at $x = 4$.

21. $f(-2) = \frac{3}{-2 + 2} = \frac{3}{0}$

Therefore, the function is

not continuous at $x = -2$.

23. $f(1) = 4 - 1 = 3$

$$\lim_{x \to 1^-} f(x) = (1)^2 + 2 = 3$$

$$\lim_{x \to 1^+} f(x) = 4 - 1 = 3$$

Therefore, $\lim_{x \to 1} f(x) = 3$, and

$\lim_{x \to 1} f(x) = f(1)$ so the function is

continuous at $x = 1$.

25. $\lim_{x \to 2^+} f(x) = 3(2) - 1 = 5$

so $\lim_{x \to 2^+} f(x)$ must also equal 5.

$$\lim_{x \to 2^+} f(x) = \lim_{x \to 2^+} (2x + a) = 4 + a$$

$4 + a = 5$

$a = 1$

27. $h(2) = 2^2 - 4 = 0$

$$\lim_{x \to -2^-} h(x) = \lim_{x \to -2} (x^2 - 4) = (-2)^2 - 4 = 0$$

$$\lim_{x \to -2^+} h(x) = \lim_{x \to -2} (2 - |x|) = 2 - 2 = 0$$

Therefore, $\lim_{x \to -2} h(x) = 0$.

$\lim_{x \to -2} h(x) = h(-2)$ so the function is

continuous at $x = -2$.

29. $P(x) = \begin{cases} 29, & 0 < x \le 1 \\ 23[\![x]\!] + 29, & x > 1 \end{cases}$

$\lim\limits_{x \to 1^-} P(x) = 29$

$\lim\limits_{x \to 1^+} P(x) = 23 + 29 = 52$

Therefore, $P(x)$ is not continuous at $x = 1$.

$\lim\limits_{x \to 2^-} P(x) = 23 + 29 = 52$

$\lim\limits_{x \to 2^+} P(x) = 23(2) + 29 = 75$

Therefore, $P(x)$ is not continuous at $x = 2$.

$\lim\limits_{x \to 3^-} P(x) = 23(2) + 29 = 75$

$\lim\limits_{x \to 3^+} P(x) = 23(3) + 29 = 98$

Therefore, $P(x)$ is not continuous at $x = 3$.

$\lim\limits_{x \to 6.5} P(x) = 23[\![6.5]\!] + 29 = 167$

$\qquad = P(6.5)$

The function is continuous at $x = 6.5$.

31. $\lim\limits_{r \to n} k = \lim\limits_{r \to n} \dfrac{r}{(n-r)c}$

$\qquad = \dfrac{n}{(n-n)c}$

The limit does not exist. The function is not defined at $r = n$.

Section 2.4   (pages 104-107)

1. $\lim\limits_{x \to 0^-} \dfrac{1}{x} = -\infty$

$\lim\limits_{x \to 0^+} \dfrac{1}{x} = +\infty$

vertical asymptote: $x = 0$

3. $\lim\limits_{x \to -3^-} \dfrac{1}{x+3} = -\infty$

$\lim\limits_{x \to -3^+} \dfrac{1}{x+3} = +\infty$

vertical asymptote: $x = -3$

5. $\lim\limits_{x \to -1^-} \dfrac{-2}{x+1} = +\infty$

$\lim\limits_{x \to -1^+} \dfrac{-2}{x+1} = -\infty$

vertical asymptote: $x = -1$

7. $f(x) = \dfrac{1}{x(x+3)}$

$\lim\limits_{x \to -3^-} f(x) = +\infty$

$\lim\limits_{x \to -3^+} f(x) = -\infty$

$\lim\limits_{x \to 0^-} f(x) = -\infty$

$\lim\limits_{x \to 0^+} f(x) = +\infty$

vertical asymptotes: $x = -3, x = 0$

9.　$\displaystyle\lim_{x\to-\infty}\frac{1}{x}=0$

　　$\displaystyle\lim_{x\to+\infty}\frac{1}{x}=0$

　　horizontal asymptote: $y=0$

11.　Divide numerator and denominator by x:

$$\lim_{x\to-\infty}\frac{x}{x+3}=\lim_{x\to-\infty}\frac{1}{1+\frac{3}{x}}=\frac{1}{1+0}=1$$

$$\lim_{x\to+\infty}\frac{x}{x+3}=\lim_{x\to+\infty}\frac{1}{1+\frac{3}{x}}=\frac{1}{1+0}=1$$

horizontal asymptote: $y=1$

13.　Divide numerator and denominator by x:

$$\lim_{x\to-\infty}\frac{x+1}{2x-3}=\lim_{x\to-\infty}\frac{1+\frac{1}{x}}{2-\frac{3}{x}}=\frac{1+0}{2-0}=\frac{1}{2}$$

$$\lim_{x\to+\infty}\frac{x+1}{2x-3}=\lim_{x\to+\infty}\frac{1+\frac{1}{x}}{2-\frac{3}{x}}=\frac{1+0}{2-0}=\frac{1}{2}$$

horizontal asymptote: $y=\frac{1}{2}$

15.　Divide numerator and denominator by x^2:

$$\lim_{x\to-\infty}\frac{x-2}{x^2+1}=\lim_{x\to-\infty}\frac{\frac{1}{x}-\frac{2}{x^2}}{1+\frac{1}{x^2}}=\frac{0-0}{1+0}=0$$

$$\lim_{x\to+\infty}\frac{x-2}{x^2+1}=\lim_{x\to+\infty}\frac{\frac{1}{x}-\frac{2}{x^2}}{1+\frac{1}{x^2}}=\frac{0-0}{1+0}=0$$

horizontal asymptote: $y=0$

17.　Divide numerator and denominator by x^2:

$$\lim_{x\to-\infty}\frac{x^2-x+3}{4x^2+5}=\lim_{x\to-\infty}\frac{1-\frac{1}{x}+\frac{3}{x^2}}{4+\frac{5}{x^2}}=\frac{1-0+0}{4+0}=\frac{1}{4}$$

$$\lim_{x\to+\infty}\frac{x^2-x+3}{4x^2+5}=\lim_{x\to+\infty}\frac{1-\frac{1}{x}+\frac{3}{x^2}}{4+\frac{5}{x^2}}=\frac{1-0+0}{4+0}=\frac{1}{4}$$

horizontal asymptote: $y=\frac{1}{4}$

19.　$\displaystyle\lim_{x\to1^-}\frac{2x}{x^2-1}=-\infty$

21.　$\displaystyle\lim_{x\to+\infty}\frac{120x^2-600x+3}{2x^2-10x+1}=\lim_{x\to+\infty}\frac{120-\frac{600}{x}+\frac{3}{x^2}}{2-\frac{10}{x}+\frac{1}{x^2}}=\frac{120-0+0}{2-0+0}=\frac{120}{2}=60$

　　sales: \$60,000

23.　$\displaystyle\lim_{x\to+\infty}\overline{C}(x)=\lim_{x\to+\infty}\left(10+\frac{1000}{x}\right)=10+0=10$

25a.　$N(10)=\dfrac{100+45(10)}{1+0.05(10)}\approx366$

25b.　$N(25)=\dfrac{100+45(25)}{1+0.05(25)}\approx544$

27. $\lim\limits_{t\to\infty} \dfrac{9200t^2 + 800}{t^2 + 5} = \lim\limits_{t\to\infty} \dfrac{9200 + \frac{800}{t^2}}{1 + \frac{5}{t^2}} = \dfrac{9200 + 0}{1 + 0} = 9200$

29.

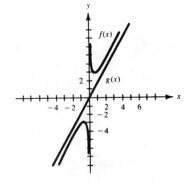

31.

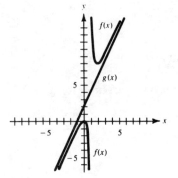

Section 2.4A (pages 113-115)

1.   vertical asymptote: $x = 5$
     horizontal asymptote: $y = 0$

3.   vertical asymptotes: $x = -1$, $x = 1$
     horizontal asymptote: $y = 0$

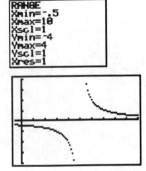

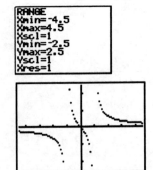

5.　vertical asymptote: $x = 1$
horizontal asymptote: $y = 0$

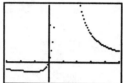

7.　no vertical asymptote
horizontal asymptote: $y = 0$

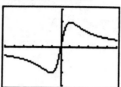

9.　no horizontal asymptote
vertical asymptote: $v = 3$

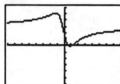

11.　No. $\lim\limits_{x \to 1^-} f(x) = -\infty$

$\lim\limits_{x \to 1^+} f(x) = +\infty$

13.　Yes. $\lim\limits_{x \to 1} f(x) = \lim\limits_{x \to 1} \dfrac{x}{x^2 + 1}$

$= \dfrac{1}{1 + 1} = \dfrac{1}{2}$

15.

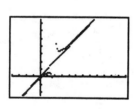

17.　Since $y = \dfrac{x^2 - 7x + 12}{x^2 - 9} = \dfrac{(x - 3)(x - 4)}{(x - 3)(x + 3)}$

$= \dfrac{x - 4}{x + 3} \qquad (x \neq 3)$

the graph of $y = \dfrac{x^2 - 7x + 12}{x^2 - 9}$

should look like the graph

of $y = \dfrac{x - 4}{x + 3}$ with a hole at $x = 3$.

19a. The vertical asymptotes are at $x = \pm\sqrt{5}$ (approximately ± 2.24).

19b.

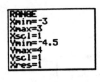
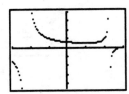

Section 2.5 (pages 123-126)

1.
$$m_{\tan} = \lim_{x \to 1} \frac{f(x) - f(1)}{x - 1} = \lim_{x \to 1} \frac{3x^2 - 3}{x - 1} = \lim_{x \to 1} \frac{3(x^2 - 1)}{x - 1} = \lim_{x \to 1} \frac{3(x + 1)(x - 1)}{x - 1}$$
$$= \lim_{x \to 1} 3(x + 1) = 6$$

3.
$$m_{\tan} = \lim_{x \to 2} \frac{f(x) - f(2)}{x - 2} = \lim_{x \to 2} \frac{\frac{1}{2}x^2 - 2}{x - 2} = \lim_{x \to 2} \frac{x^2 - 4}{2(x - 2)} = \lim_{x \to 2} \frac{(x + 2)(x - 2)}{2(x - 2)}$$
$$= \lim_{x \to 2} \frac{x + 2}{2} = 2$$

5.
$$m_{\tan} = \lim_{x \to 2} \frac{f(x) - f(2)}{x - 2} = \lim_{x \to 2} \frac{(1 - x^2) - (-3)}{x - 2} = \lim_{x \to 2} \frac{4 - x^2}{x - 2} = \lim_{x \to 2} \frac{(2 - x)(2 + x)}{x - 2}$$
$$= \lim_{x \to 2} (-1)(2 + x) = -4$$

7.
$$m_{\tan} = \lim_{x \to -1} \frac{f(x) - f(-1)}{x - (-1)} = \lim_{x \to -1} \frac{(x^2 + x) - 0}{x + 1} = \lim_{x \to -1} \frac{x(x + 1)}{x + 1}$$
$$= \lim_{x \to -1} x = -1$$

9.
$$m_{\tan} = \lim_{x \to 2} \frac{f(x) - f(2)}{x - 2} = \lim_{x \to 2} \frac{(2x^2 - x + 3) - 9}{x - 2} = \lim_{x \to 2} \frac{2x^2 - x - 6}{x - 2}$$
$$= \lim_{x \to 2} \frac{(2x + 3)(x - 2)}{x - 2} = \lim_{x \to 2} (2x + 3) = 7$$

11.
$$m_{\tan} = \lim_{x \to 1} \frac{f(x) - f(1)}{x - 1} = \lim_{x \to 1} \frac{(x^3 + 2) - 3}{x - 1} = \lim_{x \to 1} \frac{x^3 - 1}{x - 1}$$
$$= \lim_{x \to 1} \frac{(x - 1)(x^2 + x + 1)}{x - 1} = \lim_{x \to 1} (x^2 + x + 1) = 3$$

13. $m_{\tan} = \lim\limits_{x \to 1} \dfrac{f(x) - f(1)}{x - 1} = \lim\limits_{x \to 1} \dfrac{(x^2 + 2) - 3}{x - 1} = \lim\limits_{x \to 1} \dfrac{x^2 - 1}{x - 1} = \lim\limits_{x \to 1} \dfrac{(x+1)(x-1)}{x - 1}$

$\qquad = \lim\limits_{x \to 1} (x + 1) = 2 \qquad\qquad y - y_1 = m(x - x_1)$

$\qquad\qquad\qquad\qquad\qquad\qquad\qquad\; y - 3 = 2(x - 1)$

$\qquad\qquad\qquad\qquad\qquad\qquad\qquad\; y - 3 = 2x - 2$

$\qquad\qquad\qquad\qquad\qquad\qquad\qquad\qquad\;\; y = 2x + 1$

15. $m_{\tan} = \lim\limits_{x \to -1} \dfrac{f(x) - f(-1)}{x - (-1)} = \lim\limits_{x \to -1} \dfrac{(x^2 - 4x) - 5}{x + 1} = \lim\limits_{x \to -1} \dfrac{(x - 5)(x + 1)}{x + 1}$

$\qquad = \lim\limits_{x \to -1} (x - 5) = -6 \qquad\qquad y - y_1 = m(x - x_1)$

$\qquad\qquad\qquad\qquad\qquad\qquad\qquad\; y - 5 = -6[x - (-1)]$

$\qquad\qquad\qquad\qquad\qquad\qquad\qquad\; y - 5 = -6x - 6$

$\qquad\qquad\qquad\qquad\qquad\qquad\qquad\qquad\;\; y = -6x - 1$

17. $m_{\tan} = \lim\limits_{x \to -1} \dfrac{f(x) - f(-1)}{x - (-1)} = \lim\limits_{x \to -1} \dfrac{2x^3 - (-2)}{x + 1} = \lim\limits_{x \to -1} \dfrac{2(x^3 + 1)}{x + 1}$

$\qquad = \lim\limits_{x \to -1} \dfrac{2(x + 1)(x^2 - x + 1)}{x + 1} = \lim\limits_{x \to -1} 2(x^2 - x + 1) = 6$

$\qquad\qquad\qquad\qquad\qquad\qquad\qquad\; y - y_1 = m(x - x_1)$

$\qquad\qquad\qquad\qquad\qquad\qquad\qquad\; y - (-2) = 6[x - (-1)]$

$\qquad\qquad\qquad\qquad\qquad\qquad\qquad\; y + 2 = 6x + 6$

$\qquad\qquad\qquad\qquad\qquad\qquad\qquad\qquad\;\; y = 6x + 4$

19. $\dfrac{f(2) - f(0)}{2 - 0} = \dfrac{7 - 3}{2 - 0} = \dfrac{4}{2} = 2$ \qquad 21. $\dfrac{f(4) - f(1)}{4 - 1} = \dfrac{35 - 5}{4 - 1} = \dfrac{30}{3} = 10$

23. $\dfrac{f(2) - f(-2)}{2 - (-2)} = \dfrac{0 - 4}{2 + 2} = \dfrac{-4}{4} = -1$ \qquad 25. $\dfrac{f(4) - f(2)}{4 - 2} = \dfrac{68 - 10}{4 - 2} = \dfrac{58}{2} = 29$

27. $\lim\limits_{x \to -3} \dfrac{f(x) - f(-3)}{x - (-3)} = \lim\limits_{x \to -3} \dfrac{-x^2 - (-9)}{x + 3} = \lim\limits_{x \to -3} \dfrac{9 - x^2}{x + 3} = \lim\limits_{x \to -3} \dfrac{(3 + x)(3 - x)}{x + 3}$

$\qquad = \lim\limits_{x \to -3} (3 - x) = 6$

29. $\lim\limits_{x \to 1/2} \dfrac{f(x) - f\left(\frac{1}{2}\right)}{x - \frac{1}{2}} = \lim\limits_{x \to 1/2} \dfrac{2x^2 - \frac{1}{2}}{x - \frac{1}{2}} = \lim\limits_{x \to 1/2} \dfrac{4x^2 - 1}{2x - 1} = \lim\limits_{x \to 1/2} \dfrac{(2x + 1)(2x - 1)}{2x - 1}$

$\qquad = \lim\limits_{x \to 1/2} (2x + 1) = 2$

31. $\lim\limits_{x\to 1}\dfrac{f(x)-f(1)}{x-1}=\lim\limits_{x\to 1}\dfrac{(x-x^2)-0}{x-1}=\lim\limits_{x\to 1}\dfrac{-x(x-1)}{x-1}=\lim\limits_{x\to 1}(-x)=-1$

33. $\lim\limits_{x\to -1}\dfrac{f(x)-f(-1)}{x-(-1)}=\lim\limits_{x\to -1}\dfrac{\frac{1}{3}x^3-\left(-\frac{1}{3}\right)}{x+1}=\lim\limits_{x\to -1}\dfrac{\frac{1}{3}(x^3+1)}{x+1}$

$=\lim\limits_{x\to -1}\dfrac{\frac{1}{3}(x+1)(x^2-x+1)}{x+1}=\lim\limits_{x\to -1}\dfrac{1}{3}(x^2-x+1)=1$

35. $\lim\limits_{x\to 2}\dfrac{f(x)-f(2)}{x-2}=\lim\limits_{x\to 2}\dfrac{\frac{1}{x}-\frac{1}{2}}{x-2}=\lim\limits_{x\to 2}\dfrac{\frac{2}{2x}-\frac{x}{2x}}{x-2}=\lim\limits_{x\to 2}\left[\dfrac{2-x}{2x}\cdot\dfrac{1}{x-2}\right]$

$=\lim\limits_{x\to 2}\dfrac{-1}{2x}=-\dfrac{1}{4}$

37a. If $x=10$, $p=400$. If $x=20$, $p=100$. $\qquad \dfrac{100-400}{20-10}=\dfrac{-300}{10}=-30$

37b. $\lim\limits_{x\to 20}\dfrac{(500-x^2)-100}{x-20}=\lim\limits_{x\to 20}\dfrac{400-x^2}{x-20}=\lim\limits_{x\to 20}\dfrac{(20-x)(20+x)}{x-20}$

$=\lim\limits_{x\to 20}(-1)(20+x)=-40$

39. $\lim\limits_{x\to 200}\dfrac{C(x)-C(200)}{x-200}=\lim\limits_{x\to 200}\dfrac{(0.04x^2+2000)-3600}{x-200}=\lim\limits_{x\to 200}\dfrac{0.04x^2-1600}{x-200}$

$=\lim\limits_{x\to 200}\dfrac{0.04(x^2-40000)}{x-200}=\lim\limits_{x\to 200}\dfrac{0.04(x+200)(x-200)}{x-200}$

$=\lim\limits_{x\to 200}[0.04(x+200)]=16$

41a. If $x=40$, $p=6800$. If $x=50$, $p=5000$. $\qquad \dfrac{5000-6800}{50-40}=\dfrac{-1800}{10}=-180$

41b. $\lim\limits_{x\to 50}\dfrac{(10000-2x^2)-5000}{x-50}=\lim\limits_{x\to 50}\dfrac{5000-2x^2}{x-50}=\lim\limits_{x\to 50}\dfrac{-2(x^2-2500)}{x-50}$

$=\lim\limits_{x\to 50}\dfrac{-2(x+50)(x-50)}{x-50}=\lim\limits_{x\to 50}[-2(x+50)]=-200$

43a. \$2.5 million

43b. If $t=1$, $S=0.5$. If $t=4$, $S=3$. $\qquad \dfrac{3-0.5}{4-1}=\dfrac{2.5}{3}=0.8\overline{3}$ million

Approximately \$833,333

43c. Approximately $\dfrac{1}{2}$

43d. Increasing: $0<t<4$ $\qquad$ Decreasing: $t>4$

45a.   If $t = 1$, $w = 10$. If $t = 3$, $w = 20$. $\dfrac{20 - 10}{3 - 1} = \dfrac{10}{2} = 5$ lb.

45b.   If $t = 3$, $w = 20$. If $t = 5$, $w = 40$. $\dfrac{40 - 20}{5 - 3} = \dfrac{20}{2} - 10$ lb.

47. $\lim\limits_{t \to 5} \dfrac{N(t) - N(5)}{t - 5} = \lim\limits_{t \to 5} \dfrac{[85 + (t - 10)^2] - 110}{t - 5} = \lim\limits_{t \to 5} \dfrac{(t - 10)^2 - 25}{t - 5}$

$\quad = \lim\limits_{t \to 5} \dfrac{[(t - 10) + 5][(t - 10) - 5]}{t - 5} = \lim\limits_{t \to 5} \dfrac{(t - 5)(t - 15)}{t - 5} = \lim\limits_{t \to 5} (t - 15) = -10$

(decreasing rate of change)

49a. $s = 240 - 16(0)^2 = 240$ ft.   49b. $s = 240 - 16(1)^2 = 224$ ft.

49c. $s = 240 - 16(2)^2 = 176$ ft.   49d. $\dfrac{176 - 240}{2 - 0} = \dfrac{-64}{2} = -32$ ft./sec.

49e. $\lim\limits_{t \to 1} \dfrac{(240 - 16t^2) - 224}{t - 1} = \lim\limits_{t \to 1} \dfrac{16 - 16t^2}{t - 1} = \lim\limits_{t \to 1} \dfrac{-16(t^2 - 1)}{t - 1}$

$\quad \lim\limits_{t \to 1} \dfrac{-16(t + 1)(t - 1)}{t - 1} = \lim\limits_{t \to 1} (-16)(t + 1) = -32$ ft./sec.

Section 2.5A   (page 132)

1. $m = 3$ 3. $m = 7$ 5. $m = 3$

7. $m = 3$ 9. $m = -4$ 11. $m = 5$
      $b = -4$ $b = 16$ $b = -2$
      $y = 3x - 4$ $y = -4x + 16$ $y = 5x - 2$

13.   The instantaneous rate of change   15.   The instantaneous rate of change
      is approximately 0 (the slope of   is approximately 4 (the slope of
      the tangent line).   the tangent line).

17.   The function is not defined at $x = 0$.

Section 2.6   (pages 141-145)

1. $f'(x) = \lim\limits_{h \to 0} \dfrac{3(x + h) - 3x}{h} = \lim\limits_{h \to 0} \dfrac{3x + 3h - 3x}{h} = \lim\limits_{h \to 0} \dfrac{3h}{h} = \lim\limits_{h \to 0} 3 = 3$

3. $f'(x) = \lim\limits_{h \to 0} \dfrac{[4 - (x + h)] - [4 - x]}{h} = \lim\limits_{h \to 0} \dfrac{4 - x - h - 4 + x}{h} = \lim\limits_{h \to 0} \dfrac{-h}{h}$

$\quad = \lim\limits_{h \to 0} (-1) = -1$

5. $f'(x) = \lim\limits_{h \to 0} \dfrac{[4 - (x + h)^2] - [4 - x^2]}{h} = \lim\limits_{h \to 0} \dfrac{4 - x^2 - 2xh - h^2 - 4 + x^2}{h}$

$\quad = \lim\limits_{h \to 0} \dfrac{-2xh - h^2}{h} = \lim\limits_{h \to 0} \dfrac{h(-2x - h)}{h} = \lim\limits_{h \to 0} (-2x - h) = -2x - 0 = -2x$

7. $f'(x) = \lim\limits_{h \to 0} \dfrac{[(x+h)^2 + (x+h)] - [x^2 + x]}{h} = \lim\limits_{h \to 0} \dfrac{x^2 + 2xh + h^2 + x + h - x^2 - x}{h}$

$= \lim\limits_{h \to 0} \dfrac{2xh + h^2 + h}{h} = \lim\limits_{h \to 0} \dfrac{h(2x + h + 1)}{h}$

$= \lim\limits_{h \to 0} (2x + h + 1) = 2x + 0 + 1 = 2x + 1$

9. $f'(x) = \lim\limits_{h \to 0} \dfrac{[2(x+h)^2 + 3(x+h) - 1] - [2x^2 + 3x - 1]}{h}$

$= \lim\limits_{h \to 0} \dfrac{2x^2 + 4xh + h^2 + 3x + 3h - 1 - 2x^2 - 3x + 1}{h} = \lim\limits_{h \to 0} \dfrac{4xh + h^2 + 3h}{h}$

$= \lim\limits_{h \to 0} \dfrac{h(4x + h + 3)}{h} = \lim\limits_{h \to 0} (4x + h + 3) = 4x + 0 + 3 = 4x + 3$

11. $f'(x) = \lim\limits_{h \to 0} \dfrac{(x+h)^3 - x^3}{h} = \lim\limits_{h \to 0} \dfrac{x^3 + 3x^2h + 3xh^2 + h^3 - x^3}{h}$

$= \lim\limits_{h \to 0} \dfrac{3x^2h + 3xh^2 + h^3}{h} = \lim\limits_{h \to 0} \dfrac{h(3x^3 + 3xh + h^2)}{h}$

$= \lim\limits_{h \to 0} (3x^2 + 3xh + h^2) = 3x^2 + 0 + 0 = 3x^2$

13. $f'(x) = \lim\limits_{h \to 0} \dfrac{3(x+h)^2 - 3x^2}{h} = \lim\limits_{h \to 0} \dfrac{3x^2 + 6xh + 3h^2 - 3x^2}{h} = \lim\limits_{h \to 0} \dfrac{6xh + 3h^2}{h}$

$= \lim\limits_{h \to 0} \dfrac{h(6x + 3h)}{h} = \lim\limits_{h \to 0} (6x + 3h) = 6x + 0 = 6x$

13a. $f'(1) = 6(1) = 6$ 13b. $f'(3) = 6(3) = 18$

13c. $f'(-1) = 6(-1) = -6$

15. $\dfrac{dy}{dx} = \lim\limits_{h \to 0} \dfrac{[(x+h)^2 + 3(x+h)] - [x^2 + 3x]}{h}$

$= \lim\limits_{h \to 0} \dfrac{x^2 + 2xh + h^2 + 3x + 3h - x^2 - 3x}{h} = \lim\limits_{h \to 0} \dfrac{2xh + h^2 + 3h}{h}$

$= \lim\limits_{h \to 0} \dfrac{h(2x + h + 3)}{h} = \lim\limits_{h \to 0} (2x + h + 3) = 2x + 0 + 3 = 2x + 3$

15a. $\dfrac{dy}{dx}\bigg|_{x\,=\,-1} = 2(-1) + 3 = 1$ 15b. $\dfrac{dy}{dx}\bigg|_{x\,=\,2} = 2(2) + 3 = 7$

15c. $\dfrac{dy}{dx}\bigg|_{x\,=\,3} = 2(3) + 3 = 9$

17. $f'(x) = \lim\limits_{h \to 0} \dfrac{4(x+h)^2 - 4x^2}{h} = \lim\limits_{h \to 0} \dfrac{4x^2 + 8xh + 4h^2 - 4x^2}{h} = \lim\limits_{h \to 0} \dfrac{8xh + 4h^2}{h}$

$\qquad = \lim\limits_{h \to 0} \dfrac{h(8x + 4h)}{h} = \lim\limits_{h \to 0} (8x + 4h) = 8x + 0 = 8x$

$$f'(1) = 8(1) = 8$$
$$y - 4 = 8(x - 1)$$
$$y - 4 = 8x - 8$$
$$y = 8x - 4$$

19. $f'(x) = \lim\limits_{h \to 0} \dfrac{[(x+h)^2 - (x+h) - 6] - [x^2 - x - 6]}{h}$

$\qquad = \lim\limits_{h \to 0} \dfrac{x^2 + 2xh + h^2 - x - h - 6 - x^2 + x + 6}{h} = \lim\limits_{h \to 0} \dfrac{2xh + h^2 - h}{h}$

$\qquad = \lim\limits_{h \to 0} \dfrac{h(2x + h - 1)}{h} = \lim\limits_{h \to 0} (2x + h - 1) = 2x + 0 - 1 = 2x - 1$

$$f'(3) = 2(3) - 1 = 5$$
$$y - 0 = 5(x - 3)$$
$$y = 5x - 15$$

21. $f'(x) = \lim\limits_{h \to 0} \dfrac{\frac{1}{(x+h)+2} - \frac{1}{x+2}}{h} = \lim\limits_{h \to 0} \dfrac{1}{h}\left[\dfrac{x+2}{(x+h+2)(x+2)} - \dfrac{x+h+2}{(x+2)(x+h+2)}\right]$

$\qquad = \lim\limits_{h \to 0} \dfrac{x+2-x-h-2}{h(x+h+2)(x+2)} = \lim\limits_{h \to 0} \dfrac{-h}{h(x+h+2)(x+2)}$

$\qquad = \lim\limits_{h \to 0} \dfrac{-1}{(x+h+2)(x+2)} = \dfrac{-1}{(x+0+2)(x+2)} = \dfrac{-1}{(x+2)^2}$

$$f'(-1) = \dfrac{-1}{(-1+2)^2} = -1$$
$$y - 1 = -1[x - (-1)]$$
$$y - 1 = -x - 1$$
$$y = -x$$

23a. $f'(x) = \lim\limits_{h \to 0} \dfrac{\frac{1}{x+h} - \frac{1}{x}}{h} = \lim\limits_{h \to 0} \dfrac{1}{h}\left[\dfrac{x}{(x+h)x} - \dfrac{x+h}{x(x+h)}\right] = \lim\limits_{h \to 0} \dfrac{x-x-h}{h(x+h)x}$

$\qquad = \lim\limits_{h \to 0} \dfrac{-h}{h(x+h)x} = \lim\limits_{h \to 0} \dfrac{-1}{(x+h)x} = \dfrac{-1}{(x+0)x} = \dfrac{-1}{x^2}$

23b. $f(0)$ does not exist

25a. $y' = \lim\limits_{h \to 0} \dfrac{[m(x+h)+b] - [mx+b]}{h} = \lim\limits_{h \to 0} \dfrac{mx + mh + b - mx - b}{h}$

$\qquad = \lim\limits_{h \to 0} \dfrac{mh}{h} = \lim\limits_{h \to 0} m = m$

25b. Yes, the answers check.

27a.

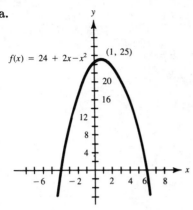

$f(x) = 24 + 2x - x^2$

(1, 25)

27b. When $f'(x) = 0$ the tangent line is horizontal: $(1, 25)$

27c. When $f'(x) < 0$ the tangent line is falling: $x > 1$

27d. When $f'(x) > 0$ the tangent line is rising: $x < 1$

29a.
$$f'(x) = \lim_{h \to 0} \frac{[4 - (x+h)^2] - [4 - x^2]}{h} = \lim_{h \to 0} \frac{4 - x^2 - 2xh - h^2 - h^2 - 4 + x^2}{h}$$

$$= \lim_{h \to 0} \frac{-2xh - h^2}{h} = \lim_{h \to 0} \frac{h(-2x - h)}{h} = \lim_{h \to 0} (-2x - h) = -2x - 0 = -2x$$

$$f'(1) = -2(1) = -2$$
$$y - 3 = -2(x - 1)$$
$$y - 3 = -2x + 2$$
$$y = -2x + 5$$

29b. The normal line must have slope $\frac{1}{2}$.

$$y - 3 = \tfrac{1}{2}(x - 1)$$
$$y - 3 = \tfrac{1}{2}x - \tfrac{1}{2}$$
$$y = \tfrac{1}{2}x + \tfrac{5}{2}$$

31a. 1   **31b.** 1   **31c.** $-\frac{3}{2}$ **31d.** 0

33a.
$$\frac{dC}{dx} = \lim_{h \to 0} \frac{[900 + 0.8(x+h)^2] - [900 + 0.8x^2]}{h}$$

$$= \lim_{h \to 0} \frac{900 + 0.8x^2 + 1.6xh + 0.8h^2 - 900 - 0.8x^2}{h} = \lim_{h \to 0} \frac{1.6xh + 0.8h^2}{h}$$

$$= \lim_{h \to 0} \frac{h(1.6x + 0.8h)}{h} = \lim_{h \to 0} (1.6x + 0.8h) = 1.6x + 0 = 1.6x$$

33b. $\left. \dfrac{dC}{dx} \right|_{x=5} = 1.6(5) = 8$

33c. $C = 900 + 0.8(5)^2 = 920$

35.
$$P'(x) = \lim_{h \to 0} \frac{[40(x+h) - (x+h)^2 - 300] - [40x - x^2 - 300]}{h}$$

$$= \lim_{h \to 0} \frac{40x + 40h - x^2 - 2xh - h^2 - 300 - 40x + x^2 + 300}{h}$$

$$= \lim_{h \to 0} \frac{40h - 2xh - h^2}{h} = \lim_{h \to 0} \frac{h(40 - 2x - h)}{h} = \lim_{h \to 0} (40 - 2x - h)$$

$$= 40 - 2x - 0 = 40 - 2x \qquad\qquad 40 - 2x = 0$$

$$-2x = -40$$

$$x = 20$$

37.
$$f'(t) = \lim_{h \to 0} \frac{5\sqrt{t+h} - 5\sqrt{t}}{h} = \lim_{h \to 0} \frac{5(\sqrt{t+h} - \sqrt{t})}{h} \cdot \frac{\sqrt{t+h} + \sqrt{t}}{\sqrt{t+h} + \sqrt{t}}$$

$$= \lim_{h \to 0} \frac{5(t+h-t)}{h(\sqrt{t+h} + \sqrt{t})} = \lim_{h \to 0} \frac{5h}{h(\sqrt{t+h} + \sqrt{t})} = \lim_{h \to 0} \frac{5}{\sqrt{t+h} + \sqrt{t}}$$

$$= \lim_{h \to 0} \frac{5}{\sqrt{t+0} + \sqrt{t}} = \frac{5}{2\sqrt{t}}$$

$$f'(9) = \frac{5}{2\sqrt{9}} = \frac{5}{6}$$

39. It is increasing at a decreasing rate since the slope of the tangent lines are decreasing as t increases.

Chapter 2 Review (pages 146-147)

1a. $x - 1 \geq 0$

$x \geq 1$

1b. $x + 4 \neq 0$

$x \neq -4$

2a. $f(0) = 2(0)^2 - 0 - 1 = -1$

2b. $f(-2) = 2(-2)^2 - (-2) - 1 = 9$

3.

$y = |x|$

4. $\lim_{x \to 6} (1 - x^2) = 1 - (6)^2 = -35$

5. $\lim_{x \to 0} \frac{x}{x+5} = \frac{0}{0+5} = 0$

6. $\lim_{x \to \infty} \frac{4x}{2x-7} = \lim_{x \to \infty} \frac{4}{2 - \frac{7}{x}} = \frac{4}{2-0} = 2$

7. $\lim_{x \to 1^+} \sqrt{x^2 - 1} = \sqrt{1 - 1} = 0$

8. $\lim_{x \to 1^-} [\![x]\!] = 0$

$\lim_{x \to 1^+} [\![x]\!] = 1$

Since $\lim_{x \to 1^-} [\![x]\!] \neq \lim_{x \to 1^+} [\![x]\!]$ then

$\lim_{x \to 1} [\![x]\!]$ does not exist.

9. $f(2) = 2(2)^2 - 5(2) + 1 = -1$

$\lim_{x \to 2} (2x^2 - 5x + 1) = -1$

$\lim_{x \to 2} f(x) = f(2)$

It is continuous at $x = 2$.

10. $f(2) = \dfrac{10(2)}{2-2} = \dfrac{20}{0}$ It is not continuous at $x = 2$.

11. $f(2) = -2-1 = -3$

$\lim\limits_{x\to 2^-} f(x) = -2-1 = -3$

$\lim\limits_{x\to 2^+} f(x) = (2)^2 - 7 = -3$

so $\lim\limits_{x\to 2} f(x) = -3$

$\lim\limits_{x\to 2} f(x) = f(2)$

It is continuous at $x = 2$.

12. $f(2) = -2-1 = -3$

$\lim\limits_{x\to 2^-} f(x) = -2-1 = -3$

$\lim\limits_{x\to 2^+} f(x) = (2)^2 - 9 = -5$

$\lim\limits_{x\to 2} f(x)$ does not exist.

It is not continuous at $x = 2$.

13a. $\lim\limits_{x\to 3^-} \dfrac{x}{x-3} = -\infty$ $\lim\limits_{x\to 3^+} \dfrac{x}{x-3} = +\infty$ vertical asymptote: $x = 3$

13b. $\lim\limits_{x\to -\infty} \dfrac{x}{x-3} = \lim\limits_{x\to -\infty} \dfrac{1}{1 - \frac{3}{x}} = \dfrac{1}{1-0} = 1$

$\lim\limits_{x\to +\infty} \dfrac{x}{x-3} = \lim\limits_{x\to +\infty} \dfrac{1}{1 - \frac{3}{x}} = \dfrac{1}{1-0} = 1$ horizontal asymptote: $y = 1$

14a. $\lim\limits_{x\to 3^-} \dfrac{1}{x-3} = -\infty$ $\lim\limits_{x\to 3^+} \dfrac{1}{x-3} = +\infty$ vertical asymptote: $x = 3$

14b. $\lim\limits_{x\to -\infty} \dfrac{1}{x-3} = \lim\limits_{x\to -\infty} \dfrac{\frac{1}{x}}{1 - \frac{3}{x}} = \dfrac{0}{1-0} = 0$

$\lim\limits_{x\to +\infty} \dfrac{1}{x-3} = \lim\limits_{x\to +\infty} \dfrac{\frac{1}{x}}{1 - \frac{3}{x}} = \dfrac{0}{1-0} = 0$ horizontal asymptote: $y = 0$

15a. $\lim\limits_{x\to -3^-} \dfrac{x}{x^2-9} = -\infty$ $\lim\limits_{x\to -3^+} \dfrac{x}{x^2-9} = +\infty$ vertical asymptote: $x = -3$

$\lim\limits_{x\to 3^-} \dfrac{x}{x^2-9} = -\infty$ $\lim\limits_{x\to 3^+} \dfrac{x}{x^2-9} = +\infty$ vertical asymptote: $x = 3$

15b. $\lim\limits_{x\to -\infty} \dfrac{x}{x^2-9} = \lim\limits_{x\to -\infty} \dfrac{\frac{1}{x}}{1 - \frac{9}{x^2}} = \dfrac{0}{1-0} = 0$

$\lim\limits_{x\to +\infty} \dfrac{x}{x^2-9} = \lim\limits_{x\to +\infty} \dfrac{\frac{1}{x}}{1 - \frac{9}{x^2}} = \dfrac{0}{1-0} = 0$ horizontal asymptote: $y = 0$

16a. $\lim\limits_{x\to -\sqrt{2}^-} \dfrac{2x^2-1}{x^2-2} = +\infty$ $\lim\limits_{x\to -\sqrt{2}^+} \dfrac{2x^2-1}{x^2-2} = -\infty$ vertical asymptote: $x = -\sqrt{2}$

$\lim\limits_{x\to \sqrt{2}^-} \dfrac{2x^2-1}{x^2-2} = -\infty$ $\lim\limits_{x\to \sqrt{2}^+} \dfrac{2x^2-1}{x^2-2} = +\infty$ vertical asymptote: $x = \sqrt{2}$

16b. $\displaystyle\lim_{x \to -\infty} \frac{2x^2 - 1}{x^2 - 2} = \lim_{x \to -\infty} \frac{2 - \frac{1}{x^2}}{1 - \frac{2}{x^2}} = \frac{2 - 0}{1 - 0} = 2$

$\displaystyle\lim_{x \to +\infty} \frac{2x^2 - 1}{x^2 - 2} = \lim_{x \to +\infty} \frac{2 - \frac{1}{x^2}}{1 - \frac{2}{x^2}} = \frac{2 - 0}{1 - 0} = 2$ horizontal asymptote: $y = 2$

17. $\displaystyle\lim_{x \to +\infty} \frac{160x^2 - 800x + 4}{x^2 - 5x + 1} = \lim_{x \to +\infty} \frac{160 - \frac{800}{x} + \frac{4}{x^2}}{1 - \frac{5}{x} + \frac{1}{x^2}} = \frac{160 - 0 + 0}{1 - 0 + 0} = 160$

Sales: \$160,000

18a. $\displaystyle m_{\text{tan}} = \lim_{x \to -1} \frac{f(x) - f(-1)}{x - (-1)} = \lim_{x \to -1} \frac{(4 - x^2) - 3}{x + 1} = \lim_{x \to -1} \frac{1 - x^2}{x + 1}$

$\displaystyle = \lim_{x \to -1} \frac{(1 + x)(1 - x)}{x + 1} = \lim_{x \to -1} (1 - x) = 2$

18b. $y - 3 = 2[x - (-1)]$

$y - 3 = 2x + 2$

$y = 2x + 5$

19. $\displaystyle\frac{f(5) - f(0)}{5 - 0} = \frac{76 - 1}{5 - 0} = \frac{75}{5} = 15$

20. $\displaystyle\lim_{x \to 3} \frac{f(x) - f(3)}{x - 3} = \lim_{x \to 3} \frac{(2x - x^2) - (-3)}{x - 3} = \lim_{x \to 3} \frac{-1(x^2 - 2x - 3)}{x - 3}$

$\displaystyle = \lim_{x \to 3} \frac{-1(x - 3)(x + 1)}{x - 3} = \lim_{x \to 3} (-1)(x + 1) = -4$

21a. When $x = 10$, $p = 370$. When $x = 20$, $p = 280$. $\displaystyle\frac{280 - 370}{20 - 10} = \frac{-90}{10} = -9$

21b. $\displaystyle\lim_{x \to 10} \frac{(400 - 0.3x^2) - 370}{x - 10} = \lim_{x \to 10} \frac{30 - 0.3x^2}{x - 10} = \lim_{x \to 10} \frac{-0.3(x^2 - 100)}{x - 10}$

$\displaystyle\lim_{x \to 10} \frac{-0.3(x + 10)(x - 10)}{x - 10} = \lim_{x \to 10} (-0.3)(x + 10) = -6$

22a. $R(100) = 3(100) - 0.02(100)^2 = \$100$

22b. $\displaystyle\frac{R(120) - R(100)}{120 - 100} = \frac{72 - 100}{120 - 100} = \frac{-28}{20} = -1.4$

22c. $R'(x) = \lim\limits_{h \to 0} \dfrac{[3(x+h) - 0.02(x+h)^2] - [3x - 0.02x^2]}{h}$

$= \lim\limits_{h \to 0} \dfrac{3x + 3h - 0.02x^2 - 0.04xh - 0.02h^2 - 3x + 0.02x^2}{h}$

$= \lim\limits_{h \to 0} \dfrac{3h - 0.04xh - 0.02h^2}{h} = \lim\limits_{h \to 0} \dfrac{h(3 - 0.04x - 0.02h)}{h}$

$= \lim\limits_{h \to 0} (3 - 0.04x - 0.02h) = 3 - 0.04x - 0 = 3 - 0.04x$

22d. $R'(120) = 3 - 0.04(120) = -1.8$

23.

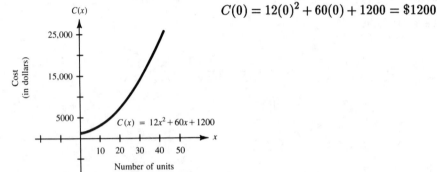

$C(0) = 12(0)^2 + 60(0) + 1200 = \$1200$

Chapter 2 Test   (pages 148-150)

1. $g(3) = 5 - 3 = 2$

$f[g(3)] = f(2) = 3(2)^2 = 12$ (b)

2. $\lim\limits_{x \to 4} (2x^2 + 1) = 2(4)^2 + 1 = 33$ (a)

3. $2x - 1 \geq 0$

$x \geq \frac{1}{2}$ (d)

4. $\lim\limits_{x \to -2} \dfrac{x^2 - 4}{3x + 6} = \lim\limits_{x \to -2} \dfrac{(x+2)(x-2)}{3(x+2)}$

$= \lim\limits_{x \to -2} \dfrac{x - 2}{3} = \dfrac{-4}{3}$ (c)

5. $\lim\limits_{x \to 2^-} f(x) = 2 + 1 = 3$

$\lim\limits_{x \to 2^+} f(x) = (2)^2 - 1 = 3$

The function is therefore continuous for all x.   (c)

6. $m_{\tan} = \lim\limits_{x\to 2} \dfrac{f(x) - f(2)}{x - 2} = \lim\limits_{x\to 2} \dfrac{\frac{2}{x-1} - 2}{x - 2} = \lim\limits_{x\to 2} \dfrac{\frac{2}{x-1} - \frac{2(x-1)}{x-1}}{x - 2}$

$\quad = \lim\limits_{x\to 2} \dfrac{\frac{2 - 2x + 2}{x-1}}{x - 2} = \lim\limits_{x\to 2} \dfrac{4 - 2x}{(x-1)(x-2)} = \lim\limits_{x\to 2} \dfrac{-2(x-2)}{(x-1)(x-2)}$

$\quad = \lim\limits_{x\to 2} \dfrac{-2}{x-1} = -2 \qquad\qquad y - 2 = -2(x - 2)$

$\qquad\qquad\qquad\qquad\qquad\qquad y - 2 = -2x + 4$

$\qquad\qquad\qquad\qquad\qquad\qquad\quad y = -2x + 6 \qquad\qquad$ (b)

7. $f(-1) = -(-1)^2 - 2(-1) + 1 = 2 \qquad\qquad$ (d)

8. The parabola opens downward with y-intercept -1 $\qquad\qquad$ (b)

9. $\lim\limits_{x\to 4} \dfrac{f(x) - f(4)}{x - 4} = \lim\limits_{x\to 4} \dfrac{-3x^2 - (-48)}{x - 4} = \lim\limits_{x\to 4} \dfrac{-3(x^2 - 16)}{x - 4} = \lim\limits_{x\to 4} \dfrac{-3(x + 4)(x - 4)}{x - 4}$

$\quad = \lim\limits_{x\to 4} (-3)(x + 4) = -24 \qquad\qquad$ (b)

10. $f'(x) = \lim\limits_{h\to 0} \dfrac{[5(x + h) - (x + h)^2] - [5x - x^2]}{h}$

$\quad = \lim\limits_{h\to 0} \dfrac{5x + 5h - x^2 - 2xh - h^2 - 5x + x^2}{h} = \lim\limits_{h\to 0} \dfrac{5h - 2xh - h^2}{h}$

$\quad = \lim\limits_{h\to 0} \dfrac{h(5 - 2x - h)}{h} = \lim\limits_{h\to 0} (5 - 2x - h) = 5 - 2x - 0 = 5 - 2x \qquad\qquad$ (c)

11. $\dfrac{f(3) - f(1)}{3 - 1} = \dfrac{25 - (-1)}{3 - 1} = \dfrac{26}{2} = 13$

12. $\dfrac{[(x + h)^2 + 4(x + h)] - [x^2 + 4x]}{h} = \dfrac{x^2 + 2xh + h^2 + 4x + 4h - x^2 - 4x}{h}$

$\quad = \dfrac{2xh + h^2 + 4h}{h} = \dfrac{h(2x + h + 4)}{h} = 2x + h + 4$

13a. $\dfrac{[1 - 2(x + h)^2] - [1 - 2x^2]}{h} = \dfrac{1 - 2x^2 - 4xh - 2h^2 - 1 + 2x^2}{h} = \dfrac{-4xh - 2h^2}{h}$

$\quad = \dfrac{h[-4x - 2h]}{h} = -4x - 2h$

13b. $\lim\limits_{h\to 0} (-4x - 2h) = -4x - 0 = -4x$

14. $\lim\limits_{x\to 2} \dfrac{f(x) - f(2)}{x - 2} = \lim\limits_{x\to 2} \dfrac{[4x^2 - x] - 14}{x - 2} = \lim\limits_{x\to 2} \dfrac{(x - 2)(4x + 7)}{x - 2} = \lim\limits_{x\to 2} (4x + 7) = 15$

15.
$$f'(x) = \lim_{h \to 0} \frac{[(x+h)^2 - 3(x+h)] - [x^2 - 3x]}{h}$$

$$= \lim_{h \to 0} \frac{x^2 + 2xh + h^2 - 3x - 3h - x^2 + 3x}{h} = \lim_{h \to 0} \frac{2xh + h^2 - 3h}{h}$$

$$= \lim_{h \to 0} \frac{h(2x + h - 3)}{h} = = \lim_{h \to 0} (2x + h - 3) = 2x + 0 - 3 = 2x - 3$$

$$2x - 3 = 0, \; x = \frac{3}{2}$$

$$f\left(\frac{3}{2}\right) = \left(\frac{3}{2}\right)^2 - 3\left(\frac{3}{2}\right) = \frac{9}{4} - \frac{9}{2} = \frac{9}{4} - \frac{18}{4} = \frac{-9}{4}$$

16.

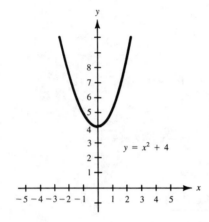

$y = x^2 + 4$

17. $C(150) = 500 + 3(150) + 0.2(150)^2 = \$5450$

18a.
$$C'(x) = \lim_{h \to 0} \frac{[0.01(x+h)^2 + 400] - [0.01x^2 + 400]}{h}$$

$$= \lim_{h \to 0} \frac{0.01x^2 + 0.02xh + 0.01h^2 + 400 - 0.01x^2 - 400}{h}$$

$$= \lim_{h \to 0} \frac{0.02xh + 0.01h^2}{h} = \lim_{h \to 0} \frac{h(0.02x + 0.01h)}{h} = \lim_{h \to 0} (0.02x + 0.01h)$$

$$= 0.02x + 0 = 0.02x$$

18b. $C'(100) = 0.02(100) = 2$

19. $\displaystyle \lim_{t \to \infty} \frac{2100t^2 + 300}{t^2 + 1} = \lim_{t \to \infty} \frac{2100 + \dfrac{300}{t^2}}{1 + \dfrac{1}{t^2}} = \frac{2100 + 0}{1 + 0} = 2100$

20. $\dfrac{s(20) - s(10)}{20 - 10} = \dfrac{-5984 - (-1344)}{20 - 10} = \dfrac{-4640}{10} = -464$ ft./sec.

CHAPTER 3: DIFFERENTIATION AND APPLICATIONS

Section 3.1 (pages 156-159)

1. $f'(x) = 2x$

3. $f'(x) = 0$

5. $\dfrac{dy}{dx} = 1$

7. $h'(x) = -6 \cdot 1$
$$= -6$$

9. $s'(t) = 3 \cdot 2t + 0$
$$= 6t$$

11. $f'(x) = 3 \cdot 2x + 9 \cdot 1 - 0$
$$= 6x + 9$$

13. $y = 3x^{1/2}$
$$\dfrac{dy}{dx} = 3 \cdot \dfrac{1}{2}x^{-1/2}$$
$$= \dfrac{3}{2\sqrt{x}}$$

15. $y = x^{1/2} + 3$
$$\dfrac{dy}{dx} = \dfrac{1}{2}x^{-1/2} + 0$$
$$= \dfrac{1}{2\sqrt{x}}$$

17. $f(x) = x^{1/3}$
$$f'(x) = \dfrac{1}{3}x^{-2/3}$$
$$= \dfrac{1}{3\sqrt[3]{x^2}}$$

19. $y = \dfrac{1}{7}x + \dfrac{5}{7}x^2$
$$\dfrac{dy}{dx} = \dfrac{1}{7} + \dfrac{5}{7} \cdot 2x$$
$$= \dfrac{1}{7} + \dfrac{10}{7}x \quad \text{or} \quad \dfrac{1 + 10x}{7}$$

21. $y = x^{-2} + x^{-3}$
$$\dfrac{dy}{dx} = -2x^{-3} + (-3)x^{-4}$$
$$= \dfrac{-2}{x^2} - \dfrac{3}{x^4}$$

23. $y = \dfrac{1}{2}x^2 + \dfrac{1}{3}x^3$
$$\dfrac{dy}{dx} = \dfrac{1}{2} \cdot 2x + \dfrac{1}{3} \cdot 3x^2$$
$$= x + x^2$$

25. $f'(x) = 2\left(\dfrac{-3}{2}\right)x^{-5/2} - 4\left(\dfrac{1}{2}\right)x^{-1/2}$
$$= -3x^{-5/2} - 2x^{-1/2}$$

27. $f(x) = 4x^2 - 28x + 49$
$$f'(x) = 4 \cdot 2x - 28 \cdot 1 + 0$$
$$= 8x - 28$$

29. $y = x^{2/3} - 5x^{-1/2}$
$$\dfrac{dy}{dx} = \dfrac{2}{3}x^{-1/3} - 5\left(\dfrac{-1}{2}\right)x^{-3/2}$$
$$= \dfrac{2}{3\sqrt[3]{x}} + \dfrac{5}{2\sqrt{x^3}}$$

31. $f'(x) = -6x$
$$f'(1) = -6(1)$$
$$= -6$$

33. $f'(x) = 6x^2 + 8x$

        $f'(-1) = 6(-1)^2 + 8(-1)$

               $= -2$

35. $f(x) = x + 2x^{-1}$

        $f'(x) = 1 - 2x^{-2}$

               $= 1 - \dfrac{2}{x^2}$

        $f'(-1) = 1 - \dfrac{2}{(-1)^2}$

               $= -1$

37. $f(x) = \frac{1}{3}x^{-1/2}$

        $f'(x) = \frac{1}{3}\left(-\frac{1}{2}\right)x^{-3/2}$

               $= \dfrac{-1}{6\sqrt{x^3}}$

        $f'(4) = \dfrac{-1}{6(\sqrt{4})^3}$

               $= \dfrac{-1}{48}$

39. $f(x) = \dfrac{x^2}{2x^4} + \dfrac{2}{2x^4}$

              $= \dfrac{1}{2x^2} + \dfrac{1}{x^4}$

              $= \frac{1}{2}x^{-2} + x^{-4}$

        $f'(x) = -x^{-3} - 4x^{-5}$

              $= \dfrac{-1}{x^3} - \dfrac{4}{x^5}$

        $f'(-1) = \dfrac{-1}{(-1)^3} - \dfrac{4}{(-1)^5}$

              $= 5$

41. $f'(x) = 2x$

        $f'(1) = 2(1) = 2$

        $y - y_1 = m(x - x_1)$

        $y - 1 = 2(x - 1)$

        $y - 1 = 2x - 2$

           $y = 2x - 1$

43. $f'(x) = -3x^2$

        $f'(-1) = -3(-1)^2 = -3$

        $y - y_1 = m(x - x_1)$

        $y - 2 = -3[x - (-1)]$

        $y - 2 = -3x - 3$

           $y = -3x - 1$

45a. $g'(x) = 8x + 4$

45b. $g(0) = 4(0)^2 + 4(0) + 1 = 1$

45c. $g'(x) = 8(0) + 4 = 4$

45d. $8x + 4 = 0$

           $8x = -4$

             $x = -\frac{1}{2}$

45e. $g\left(-\frac{1}{2}\right) = 4\left(\frac{-1}{2}\right)^2 + 4\left(\frac{-1}{2}\right) + 1$

              $= 0$

47a.

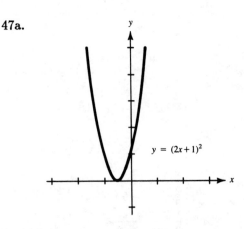

$y = (2x + 1)^2$

47b. $m = 0$

$y - y_1 = m(x - x_1)$

$y - 0 = 0\left[x - \left(-\frac{1}{2}\right)\right]$

$y = 0$

49. $f'(x) = \frac{2}{3}x^{-1/3} = \frac{2}{3\sqrt[3]{x}}$

49a.

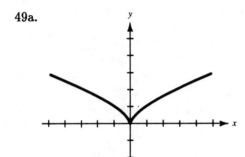

49b. It is undefined.

49c. Since the targent line is vertical, the equation is $x = 0$.

51a. $C'(x) = 0 + 600\left(\frac{2}{3}\right)x^{-1/3}$

$= \frac{400}{\sqrt[3]{x}}$ or $400x^{-1/3}$

51b. $C'(1000) = \frac{400}{\sqrt[3]{1000}} = \frac{400}{10} = 40$

53. $R'(x) = 20 - 0.06x$

$R'(100) = 20 - 0.06(100)$

$= \$14$

55. $\frac{dC}{dt} = -2t$

$\left.\frac{dC}{dt}\right|_{t=60} = -2(60) = -120$

57a. $s'(t) = -32t + 128$

57b. $v(0) = s'(0) = 128$ ft./sec.

57c. $v(3) = s'(3) = -32(3) + 128$

$= 32$ ft./sec.

Section 3.2 (pages 166-168)

1. $u(x) = x^2 \quad v(x) = 3x^2 - 1$

$f'(x) = u(x)v'(x) + u'(x)v(x)$

$= x^2(6x) + 2x(3x^2 - 1)$

$= 6x^3 + 6x^3 - 2x$

$= 12x^3 - 2x$

3. $u(x) = 1 - x^2 \quad v(x) = 1 + x^2$

$f'(x) = u(x)v'(x) + u'(x)v(x)$

$= (1 - x^2)(2x) + (-2x)(1 + x^2)$

$= 2x - 2x^3 - 2x - 2x^3$

$= -4x^3$

5. $u(x) = x^{-3}$ $v(x) = x^2 - x$

$g'(x) = u(x)v'(x) + u'(x)v(x)$

$\quad = x^{-3}(2x - 1) + (-3x^{-4})(x^2 - x)$

$\quad = 2x^{-2} - x^{-3} - 3x^{-2} + 3x^{-3}$

$\quad = 2x^{-3} - x^{-2}$

$\quad = \dfrac{2}{x^3} - \dfrac{1}{x^2} = \dfrac{2 - x}{x^3}$

7. $u(t) = t^2 + 1$ $v(t) = 2t^2 - t + 3$

$\dfrac{ds}{dt} = u(t)v'(t) + u'(t)v(t)$

$\quad = (t^2 + 1)(4t - 1) + 2t(2t^2 - t + 3)$

$\quad = 4t^3 - t^2 + 4t - 1 + 4t^3 - 2t^2 + 6t$

$\quad = 8t^3 - 3t^2 + 10t - 1$

9. $u(x) = 2$ $v(x) = x^3$

$f'(x) = \dfrac{v(x)u'(x) - u(x)v'(x)}{[v(x)]^2}$

$\quad = \dfrac{x^3(0) - 2(3x^2)}{(x^3)^2}$

$\quad = \dfrac{-6x^2}{x^6} = \dfrac{-6}{x^4}$

11. $u(x) = 2x$ $v(x) = x + 1$

$y' = \dfrac{v(x)u'(x) - u(x)v'(x)}{[v(x)]^2}$

$\quad = \dfrac{(x + 1)(2) - 2x(1)}{(x + 1)^2}$

$\quad = \dfrac{2x + 2 - 2x}{(x + 1)^2} = \dfrac{2}{(x + 1)^2}$

13. $u(x) = 5x + 1$ $v(x) = x^2 - 1$

$g'(x) = \dfrac{v(x)u'(x) - u(x)v'(x)}{[v(x)]^2}$

$\quad = \dfrac{(x^2 - 1)(5) - (5x + 1)(2x)}{(x^2 - 1)^2}$

$\quad = \dfrac{5x^2 - 5 - 10x^2 - 2x}{(x^2 - 1)^2}$

$\quad = \dfrac{-5x^2 - 2x - 5}{(x^2 - 1)^2}$

15. $u(x) = x^2$ $v(x) = x^3 + 4$

$y' = \dfrac{v(x)u'(x) - u(x)v'(x)}{[v(x)]^2}$

$\quad = \dfrac{(x^3 + 4)(2x) - x^2(3x^2)}{(x^3 + 4)^2}$

$\quad = \dfrac{2x^4 + 8x - 3x^4}{(x^3 + 4)^2}$

$\quad = \dfrac{8x - x^4}{(x^3 + 4)^2}$

17. $u(x) = x^{1/2}$ $v(x) = x + 1$

$g'(x) = \dfrac{v(x)u'(x) - u(x)v'(x)}{[v(x)]^2}$

$\quad = \dfrac{(x + 1)\left(\frac{1}{2}x^{-1/2}\right) - x^{1/2}(1)}{(x + 1)^2}$

$\quad = \dfrac{\frac{1}{2}x^{-1/2}[(x + 1) - 2x]}{(x + 1)^2}$

$\quad = \dfrac{1 - x}{2\sqrt{x}(x + 1)^2}$

19. $f(x) = \dfrac{x^2}{(x + 1)^2} = \dfrac{x^2}{x^2 + 2x + 1}$

$u(x) = x^2$ $v(x) = x^2 + 2x + 1$

$f'(x) = \dfrac{v(x)u'(x) - u(x)v'(x)}{[v(x)]^2}$

$\quad = \dfrac{(x^2 + 2x + 1)(2x) - x^2(2x + 2)}{(x^2 + 2x + 1)^2}$

$\quad = \dfrac{2x^3 + 4x^2 + 2x - 2x^3 - 2x^2}{(x^2 + 2x + 1)^2}$

$\quad = \dfrac{2x^2 + 2x}{(x^2 + 2x + 1)^2} = \dfrac{2x(x + 1)}{(x + 1)^4}$

$\quad = \dfrac{2x}{(x + 1)^3}$

21. $u(x) = x + 1 \qquad v(x) = x - 1$

$$g'(x) = \frac{v(x)u'(x) - u(x)v'(x)}{[v(x)]^2}$$

$$= \frac{(x-1)(1) - (x+1)(1)}{(x-1)^2}$$

$$= \frac{x - 1 - x - 1}{(x-1)^2}$$

$$= \frac{-2}{(x-1)^2}$$

$$g'(2) = \frac{-2}{(2-1)^2} = -2$$

23a. $u(x) = 0.2x^2 \qquad v(x) = x + 300$

$$C'(x) = u(x)v'(x) + u'(x)v(x)$$

$$= 0.2x^2(1) + 0.4x(x + 3000)$$

$$= 0.2x^2 + 0.4x^2 + 1200x$$

$$= 0.6x^2 + 1200x$$

23b. $C'(100) = 0.6(100)^2 + 1200(100)$

$$= 126,000$$

25. $u(x) = 20 \qquad v(x) = 2x + 1$

$$\frac{dp}{dx} = \frac{v(x)u'(x) - u(x)v'(x)}{[v(x)]^2}$$

$$= \frac{(2x+1)(0) - 20(2)}{(2x+1)^2}$$

$$= \frac{-40}{(2x+1)^2}$$

27a. $u(x) = 8x \qquad v(x) = 4 + x^2$

$$R'(x) = \frac{v(x)u'(x) - u(x)v'(x)}{[v(x)]^2}$$

$$= \frac{(4+x^2)(8) - 8x(2x)}{(4+x^2)^2}$$

$$= \frac{32 + 8x^2 - 16x^2}{(4+x^2)^2}$$

$$= \frac{32 - 8x^2}{(4+x^2)^2}$$

27b. $R'(2) = \frac{32 - 8(2)^2}{(4+2^2)^2}$

$$= 0$$

29. $u(x) = 500 + x \qquad v(x) = x^3 + x + 1$

$$C'(x) = \frac{(x^3 + x + 1)(1) - (500 + x)(3x^2 + 1)}{(x^3 + x + 1)^2}$$

$$= \frac{x^3 + x + 1 - 1500x^2 - 500 - 3x^3 - x}{(x^3 + x + 1)^2}$$

$$= \frac{-2x^3 - 1500x^2 - 499}{(x^3 + x + 1)^2}$$

Section 3.3   (pages 172-176)

1. $u(x) = x^2 - 3$

$g(x) = x^4$

3. $u(x) = 3x + \frac{1}{x}$

$g(x) = x^{-2}$

5. $f'(x) = 10(x + 2)^9$

7. $f'(x) = 5 \cdot 3(x^4 - 1)^2(4x^3)$

$$= 60x^3(x^4 - 1)^2$$

9. $f(x) = (5 - 3x)^{1/3}$

$f'(x) = \frac{1}{3}(5 - 3x)^{-2/3}(-3)$

$= \dfrac{-1}{\sqrt[3]{(5 - 3x)^2}}$

11. $y' = -2x^2[5(3x + 1)^4(3)] + (-4x)(3x + 1)^5$

$= -2x(3x + 1)^4[x(5)(3) + (2)(3x + 1)]$

$= -2x(3x + 1)^4(15x + 6x + 2)$

$= -2x(3x + 1)^4(21x + 2)$

13. $f'(x) = (x + 1)^3[4(2x - 3)^3(2)] + [3(x + 1)^2](2x - 3)^4$

$= (x + 1)^2(2x - 3)^3[(x + 1)(4)(2) + 3(2x - 3)]$

$= (x + 1)^2(2x - 3)^3(8x + 8 + 6x - 9)$

$= (x + 1)^2(2x - 3)^3(14x - 1)$

15. $f'(x) = \dfrac{(1 - x)^4(2) - 2x[4(1 - x)^3(-1)]}{[(1 - x)^4]^2}$

$= \dfrac{2(1 - x)^3[(1 - x) + x(4)]}{(1 - x)^8}$

$= \dfrac{2(3x + 1)}{(1 - x)^5}$

17. $y = \dfrac{5}{(x^2 + 1)^{1/2}} = 5(x^2 + 1)^{-1/2}$

$y' = 5\left(\dfrac{-1}{2}\right)(x^2 + 1)^{-3/2}(2x)$

$= \dfrac{-5x}{\sqrt{(x^2 + 1)^3}}$

19.

$$y = \frac{-3x}{(x+1)^{1/4}}$$

$$y' = \frac{(x+1)^{1/4}(-3) - (-3x)\left(\frac{1}{4}\right)(x+1)^{-3/4}(1)}{\left[(x+1)^{1/4}\right]^2}$$

$$= \frac{-3(x+1)^{-3/4}\left[(x+1) - x\left(\frac{1}{4}\right)\right]}{(x+1)^{1/2}}$$

$$= \frac{-3\left(\frac{3}{4}x + 1\right)}{(x+1)^{5/4}} = \frac{-3(3x+4)}{4\sqrt[4]{(x+1)^5}}$$

21.

$$f(x) = \frac{(x-1)^4}{(x+1)^4}$$

$$f'(x) = \frac{(x+1)^4\left[4(x-1)^3\right] - (x-1)^4\left[4(x+1)^3\right]}{\left[(x+1)^4\right]^2}$$

$$= \frac{4(x+1)^3(x-1)^3\left[(x+1) - (x-1)\right]}{(x+1)^8}$$

$$= \frac{4(x+1)^3(x-1)^3(2)}{(x+1)^8} = \frac{8(x-1)^3}{(x+1)^5}$$

23.

$$y = (u+1)^{-1} \qquad u = x^{1/2} \qquad \frac{dy}{dx} = \frac{dy}{du} \cdot \frac{du}{dx}$$

$$\frac{dy}{du} = -1(u+1)^{-2} \qquad \frac{du}{dx} = \frac{1}{2}x^{-1/2} \qquad = \frac{-1}{(u+1)^2} \cdot \frac{1}{2\sqrt{x}}$$

$$= \frac{-1}{(u+1)^2} \qquad\qquad = \frac{1}{2\sqrt{x}} \qquad = \frac{-1}{(\sqrt{x}+1)^2(2\sqrt{x})}$$

25.

$$\frac{dy}{du} = 6u^2 - 3 \qquad \frac{du}{dx} = 2x \qquad \frac{dy}{dx} = \frac{dy}{du} \cdot \frac{du}{dx}$$

$$= (6u^2 - 3)(2x)$$

$$= \left[6(1+x^2)^2 - 3\right](2x)$$

$$= \left[6(1 + 2x^2 + x^4) - 3\right](2x)$$

$$= 6x + 24x^3 + 12x^5$$

27. $\dfrac{dy}{du} = \dfrac{(u+1)(1) - u(1)}{(u+1)^2}$ $\dfrac{du}{dx} = 4x$ $\dfrac{dy}{dx} = \dfrac{dy}{du} \cdot \dfrac{du}{dx}$

$\qquad\quad = \dfrac{u+1-u}{(u+1)^2}$ $\qquad\qquad\qquad\qquad\qquad = \dfrac{1}{(u+1)^2} \cdot 4x$

$\qquad\quad = \dfrac{1}{(u+1)^2}$ $\qquad\qquad\qquad\qquad\qquad\quad = \dfrac{4x}{\left[(2x^2 - 1) + 1\right]^2}$

$\qquad\qquad\qquad\qquad\qquad\qquad\qquad\qquad\qquad\quad = \dfrac{4x}{\left[2x^2\right]^2} = \dfrac{4x}{4x^4} = \dfrac{1}{x^3}$

29. $f(x) = \dfrac{-2x}{(x^2 + 9)^{1/2}}$

$\qquad f'(x) = \dfrac{(x^2 + 9)^{1/2}(-2) - (-2x)\left[\left(\frac{1}{2}\right)(x^2 + 9)^{-1/2}(2x)\right]}{\left[(x^2 + 9)^{1/2}\right]^2}$

$\qquad\qquad = \dfrac{-2(x^2 + 9)^{-1/2}\left[(x^2 + 9) - x^2\right]}{x^2 + 9}$

$\qquad\qquad = \dfrac{-18}{(x^2 + 9)^{3/2}} = \dfrac{-18}{\sqrt{(x^2 + 9)^3}}$

$\qquad f'(0) = \dfrac{-18}{27} = -\dfrac{2}{3}$

31. $\dfrac{dy}{dx} = -2[6(x^2 + x + 1)^5(2x + 1)]$ 33. $f'(y) = 4(y^3 - 1)^3(3y^2)$

$\qquad\qquad = -12(2x + 1)(x^2 + x + 1)^5$ $\qquad\qquad\qquad\quad = 12y^2(y^3 - 1)^3$

$\qquad \left.\dfrac{dy}{dx}\right|_{x=0} = -12(1)(1) = -12$ $\qquad\qquad\qquad f'(1) = 12(1)(0)$

$\qquad\qquad\qquad\qquad\qquad\qquad\qquad\qquad\qquad\qquad\qquad = 0$

35a. $\dfrac{dy}{dt} = 2t$ $\dfrac{dx}{dt} = 12t^2$ $\dfrac{dy}{dx} = \dfrac{\frac{dy}{dt}}{\frac{dx}{dt}} = \dfrac{2t}{12t^2} = \dfrac{1}{6t}$

35b. $y = (2t + 1)^{1/2}$ $x = (2t)^{1/2}$

$\qquad \dfrac{dy}{dt} = \frac{1}{2}(2t + 1)^{-1/2}(2)$ $\dfrac{dx}{dt} = \frac{1}{2}(2t)^{-1/2}(2)$ $\dfrac{dy}{dx} = \dfrac{\frac{dy}{dt}}{\frac{dx}{dt}}$

$\qquad\qquad = \dfrac{1}{(2t + 1)^{1/2}}$ $\qquad\qquad\quad = \dfrac{1}{(2t)^{1/2}}$

$\qquad\qquad = \dfrac{1}{\sqrt{2t + 1}}$ $\qquad\qquad\qquad = \dfrac{1}{\sqrt{2t}}$ $\qquad\qquad\qquad\quad = \dfrac{\frac{1}{\sqrt{2t + 1}}}{\frac{1}{\sqrt{2t}}} = \dfrac{\sqrt{2t}}{\sqrt{2t + 1}}$

35c.
$$\frac{dy}{dt} = \frac{(2-t)(0) - 1(-1)}{(2-t)^2}$$
$$= \frac{1}{(2-t)^2}$$

$$x = t^{-1}$$
$$\frac{dx}{dt} = -t^{-2}$$
$$= \frac{-1}{t^2}$$

$$\frac{dy}{dx} = \frac{\frac{dy}{dt}}{\frac{dx}{dt}}$$
$$= \frac{\frac{1}{(2-t)^2}}{\frac{-1}{t^2}} = \frac{-t^2}{(2-t)^2}$$

37.
$$\frac{dx}{dp} = -0.4p \qquad\qquad \frac{dp}{dt} = 0.2t$$
$$\frac{dx}{dt} = \frac{dx}{dp} \cdot \frac{dp}{dt}$$
$$= (-0.4p)(0.2t)$$
$$= -0.08t(8 + 0.1t^2)$$
$$\left.\frac{dx}{dt}\right|_{t=6} = -0.08(6)\big[8 + 0.1(6)^2\big]$$
$$= -5.568$$

39.
$$p = \frac{128}{(2x+6)^{1/2}} = 128(2x+6)^{-1/2}$$
$$\frac{dp}{dx} = 128\left[\left(\frac{-1}{2}\right)(2x+6)^{-3/2}(2)\right]$$
$$= \frac{-128}{\sqrt{(2x+t)^3}}$$
$$\left.\frac{dp}{dx}\right|_{x=5} = \frac{-128}{\left(\sqrt{2(5)+6}\right)^3} = \frac{-128}{64} = -2$$

41a. $\dfrac{P(3) - P(1)}{3 - 1} = \dfrac{500 - 420}{3 - 1} = \dfrac{80}{2} = \$40$

41b.
$$P'(x) = 10[3(x-3)^2]$$
$$= 30(x-3)^2$$
$$P'(2) = 30(2-3)^2 = \$30$$

41c.
$$30(x-3)^2 = 0$$
$$x = 3 \qquad \text{(300 lawnmowers)}$$

43a. $N(0) = 50\sqrt{0+1} + 2400 = 2450$

43b.
$$N(t) = 50(t+1)^{1/2} + 2400$$
$$N'(t) = 50\left[\tfrac{1}{2}(t+1)^{-1/2}(1)\right]$$
$$= \frac{25}{\sqrt{t+1}}$$
$$N'(3) = \frac{25}{\sqrt{3+1}} = \frac{25}{2} = 12.5$$

45a.
$$P(x) = 400(101 - x)^{1/2}$$
$$P'(x) = 400\left[\tfrac{1}{2}(101-x)^{-1/2}(-1)\right]$$
$$= \frac{-200}{\sqrt{101 - x}}$$

45b. $P'(76) = \dfrac{-200}{\sqrt{101-76}} = \dfrac{-200}{5} = -40$

Section 3.4 (pages 178-179)

1. $f'(x) = 4x^3 + 4x$

$f''(x) = 12x^2 + 4$

3. $f(x) = (2x + 1)^{-1/2}$

$f'(x) = \frac{-1}{2}(2x + 1)^{-3/2}(2)$

$= -1(2x + 1)^{-3/2}$

$f''(x) = -1\left(\frac{-3}{2}\right)(2x + 1)^{-5/2}(2)$

$= \frac{3}{\sqrt{(2x + 1)^5}}$

5. $f'(x) = 4(3x - 2)^3(3)$

$= 12(3x - 2)^3$

$f'(x) = 12(3)(3x - 2)^2(3)$

$= 108(3x - 2)^2$

7. $f'(x) = 5(2x + 1)^4(2)$

$= 10(2x + 1)^4$

$f''(x) = 10(4)(2x + 1)^3(2)$

$= 80(2x + 1)^3$

$f''(0) = 80(1)$

$= 80$

9. $f'(x) = 3x^2 - 14x$

$f''(x) = 6x - 14$

$f''(-1) = 6(-1) - 14$

$= -20$

11. $y = x^{-1/2}$

$\frac{dy}{dx} = \frac{-1}{2}x^{-3/2}$

$\frac{d^2y}{dx^2} = \frac{-1}{2}\left(\frac{-3}{2}\right)x^{-5/2}$

$= \frac{3}{4\sqrt{x^5}}$

$\left.\frac{d^2y}{dx^2}\right|_{x=4} = \frac{3}{4(\sqrt{4})^5} = \frac{3}{128}$

13. $\frac{dy}{dx} = 4(3x + 1)^3(3)$

$= 12(3x + 1)^3$

$\frac{d^2y}{dx^2} = 12(3)(3x + 1)^2(3)$

$= 108(3x + 1)^2$

15. $f(x) = x^{-2}$

$f'(x) = -2x^{-3}$

$f''(x) = -2(-3)x^{-4} = 6x^{-4}$

$f'''(x) = 6(-4)x^{-5} = -24x^{-5}$

$f^{(4)}(x) = -24(-5)x^{-6} = \frac{120}{x^6}$

17. $\frac{dy}{dx} = \frac{(x - 2)(1) - x(1)}{(x - 2)^2}$

$= \frac{-2}{(x - 2)^2} = -2(x - 2)^{-2}$

$\frac{d^2y}{dx^2} = -2(-2)(x - 2)^{-3}$

$= \frac{4}{(x - 2)^3}$

19. $D_x y = 4x^3 - 12x^2$

$D_x^2 y = 12x^2 - 24x$

$D_x^3 y = 24x - 24$

$D_x^4 y = 24$

$D_x^5 y = 0$

21. $f(x) = x^{-1/2}$

$f'(x) = \frac{-1}{2}x^{-3/2}$

$f''(x) = \frac{-1}{2}\left(\frac{-3}{2}\right)x^{-5/2} = \frac{3}{4}x^{-5/2}$

$f'''(x) = \frac{3}{4}\left(\frac{-5}{2}\right)x^{-7/2} = \frac{-15}{8}x^{-7/2}$

$f^{(4)}(x) = \frac{-15}{8}\left(\frac{-7}{2}\right)x^{-9/2}$

$= \frac{105}{16\sqrt{x^9}}$

23. $y = \dfrac{2x}{(x-1)^{1/3}}$

$y' = \dfrac{(x-1)^{1/3}(2) - 2x\left[\frac{1}{3}(x-1)^{-2/3}\right]}{\left[(x-1)^{1/3}\right]^2}$

$= \dfrac{2(x-1)^{-2/3}\left[(x-1) - x\left(\frac{1}{3}\right)\right]}{(x-1)^{2/3}}$

$= \dfrac{2(x-1)^{-2/3}\left(\frac{2}{3}x - 1\right)}{(x-1)^{2/3}}$

$= \dfrac{2(2x-3)}{3(x-1)^{4/3}} = \dfrac{4x-6}{3(x-1)^{4/3}}$

$y'' = \dfrac{3(x-1)^{4/3}(4) - (4x-6)[3\left(\frac{4}{3}\right)(x-1)^{1/3}]}{\left[3(x-1)^{4/3}\right]^2}$

$= \dfrac{4(x-1)^{1/3}[3(x-1) - (4x-6)]}{9(x-1)^{8/3}}$

$= \dfrac{4(3-x)}{9(x-1)^{7/3}}$

$= \dfrac{4(3-x)}{9\sqrt[3]{(x-1)^7}}$

25. $f'(x) = 3x^2 - 1$

$f''(x) = 6x$

$6x = 0$

$x = 0$

27. $f(x) = x^{1/2} + x^{-1/2}$

$f'(x) = \frac{1}{2}x^{-1/2} + \left(\frac{-1}{2}\right)x^{-3/2}$

$f''(x) = \frac{1}{2}\left(\frac{-1}{2}\right)x^{-3/2} + \left(\frac{-1}{2}\right)\left(\frac{-3}{2}\right)x^{-5/2}$

$= \dfrac{-1}{4\sqrt{x^3}} + \dfrac{3}{4\sqrt{x^5}}$

$= \dfrac{-x+3}{4\sqrt{x^5}}$

$-x + 3 = 0$

$x = 3$

Section 3.5 (pages 188-194)

1a. $C'(x) = 40 - x$

1b. $C'(30) = 40 - 30$
$= \$10$

3a. $P'(x) = 0.06x^2 + 0.04$

3b. $P'(5) = 0.06(5)^2 + 0.04$
$= 1.54 \quad (\$1540)$

1c. $C(31) - C(30) = 779.50 - 770$
$$= \$9.50$$

3c. $P(6) - P(5) = 4.56 - 2.70$
$$= 1.86 \quad (\$1860)$$

5a. $R(x) = px$
$$= (40 - 0.05x)x$$
$$= 40x - 0.05x^2$$

7a. $P(201) - P(200) = 1604.98 - 1600$
$$= 4.98$$

7b.

5b. $C'(x) = 20$

5c. $P(x) = R(x) - C(x)$
$$= (40x - 0.05x^2) - (3500 + 20x)$$
$$= -0.05x^2 + 20x - 3500$$

5d. $P'(x) = -0.10x + 20$

7c.   325

9a. $\overline{C}(x) = \dfrac{\frac{200}{x} = 5.5x}{x}$
$$= \frac{200}{x^2} + 5.5$$

11a. $C'(x) = 2 - 6x + 3x^2$

11b. $\overline{C}(x) = \dfrac{2x - 3x^2 + x^3}{x}$
$$= 2 - 3x + x^2$$

9b. $C(x) = 200x^{-1} + 5.5x$
$C'(x) = -200x^{-2} + 5.5$
$$= \frac{-200}{x^2} + 5.5$$

11c. $\overline{C}'(x) = -3 + 2x$

9c. $\overline{C}(x) = 200x^{-2} + 5.5$
$\overline{C}'(x) = -400x^{-3} = \dfrac{-400}{x^3}$

13a. $R'(x) = 8 - 2x$

15. $50x = 1000 + 10x$
$40x = 1000$
$x = 25$

13b. $R'(2) = 8 - 2(2) = 4$

13c. $8 - 2x = 0, \quad x = 4$

17. $40x - x^2 = 75 + 12x$
$$0 = x^2 - 28x + 75$$
$$0 = (x - 3)(x - 25)$$
$x = 3 \qquad x = 25$

19a.   100 units or 600 units

19b.   400 units

19c.   300 units

19d.   100 units to 600 units

21a. $\dfrac{E(3) - E(1)}{3 - 1} = \dfrac{45 - 15}{3 - 1}$
$$= \frac{30}{2}$$
$$= 15$$

23a.   There is very little change in slope.

23b.   There is very little change in viscosity.

23c.   It increases rapidly and then levels off.

21b. $E'(t) = 12 + 8t - 3t^2$

$E'(1) = 17$

$E'(2) = 16$

$E'(3) = 9$

$E'(4) = -4$

$E'(5) = -23$

$E'(6) = -48$

21c. Approximately 3.7 hours since $E'(3.7)$ is close to 0.

29a. $\dfrac{s(3) - s(2)}{3 - 2} = \dfrac{560 - 512}{3 - 2}$

$= 48$ ft./sec.

29b. $v(t) = s'(t)$

$= 128 - 32t$

$v(3) = 128 - 32(3)$

$= 32$ ft./sec.

29c. $128 - 32t = 0$

$t = 4$

29d. $128 - 32t > 0$

$-32t > -128$

$t < 4$

so $0 < t < 4$

29e. $128 - 32t < 0$

$-32t < -128$

$t > 4$

so $4 < t < 10$

25. $N'(t) = \dfrac{(t + 2)(400) - (400t + 250)(1)}{(t + 2)^2}$

$= \dfrac{550}{(t + 2)^2}$

27. $V'(t) = 300\left(\dfrac{1}{3}\right)t^{-2/3}$

$= \dfrac{100}{\sqrt[3]{t^2}}$

$V'(8) = \dfrac{100}{(\sqrt[3]{8})^2} = 25$

31a. $s(t) = 5t + t^{-2}$

$v(t) = s'(t)$

$= 5 - 2t^{-3}$

$= 5 - \dfrac{2}{t^3}$

$v(2) = 4.75$ ft./sec.

31b. $a(t) = v'(t)$

$= 6t^{-4}$

$= \dfrac{6}{t^4}$

$a(1) = 6$ ft./sec.^{2}

33a. $\dfrac{dh}{dt} = -32t$

$\dfrac{dh}{dt}\bigg|_{t\,=\,1} = -32$ ft./sec. (the velocity after 1 second)

33b.
$$480 - 16t^2 = 0$$
$$t^2 = 30$$
$$t = \sqrt{30}$$
$$t \approx 5.48 \text{ sec.}$$

33c.
$$\left.\frac{dh}{dt}\right|_{t=\sqrt{30}} = -32\sqrt{30}$$
$$\approx -175.3 \text{ ft./sec.}$$

35a.
$$v(t) = s'(t)$$
$$= -32t + 64$$
$$-32t + 64 = 0$$
$$t = 2 \text{ sec.}$$

35b.
$$s(2) = -16(2)^2 + 64(2) + 112$$
$$= 176 \text{ ft.}$$

35c.
$$0 = -16t^2 + 64t + 112$$
$$0 = -16(t^2 - 4t - 7)$$
$$t = \frac{-(-4) \pm \sqrt{(-4)^2 - 4(1)(-7)}}{2(1)}$$
$$= \frac{4 \pm \sqrt{44}}{2}$$
$$= 2 \pm \sqrt{11}$$
$$t = 2 + \sqrt{11} \approx 5.32 \text{ sec.}$$

35d.
$$v(0) = -32(0) + 64$$
$$= 64 \text{ ft./sec.}$$

Section 3.6   (pages 202-205)

1.
$$1 - 3\frac{dy}{dx} + 0 = 0$$
$$-3\frac{dy}{dx} = -1$$
$$\frac{dy}{dx} = \frac{1}{3}$$

3.
$$2y\frac{dy}{dx} = 2 - 0$$
$$\frac{dy}{dx} = \frac{2}{2y}$$
$$\frac{dy}{dx} = \frac{1}{y}$$

5.
$$2x - 8y\frac{dy}{dx} = 0$$
$$-8y\frac{dy}{dx} = -2x$$
$$\frac{dy}{dx} = \frac{-2x}{-8y}$$
$$= \frac{x}{4y}$$

7.
$$3x^2 + 3y^2\frac{dy}{dx} = 0$$
$$3y^2\frac{dy}{dx} = -3x^2$$
$$\frac{dy}{dx} = \frac{-x^2}{y^2}$$

9.
$$[x^2 \cdot 2y\frac{dy}{dx} + 2x \cdot y^2] - 12y^2\frac{dy}{dx} + 3 = 0$$
$$2x^2y\frac{dy}{dx} - 12y^2\frac{dy}{dx} = -2xy^2 - 3$$
$$(2x^2y - 12y^2)\frac{dy}{dx} = -2xy^2 - 3 \qquad \frac{dy}{dx} = \frac{-2xy^2 - 3}{2x^2y - 12y^2} \quad \text{or} \quad \frac{2xy^2 + 3}{12y^2 - 2x^2y}$$

11. $\left[2x\dfrac{dy}{dx} + 2y\right] - \left[4x \cdot 2y\dfrac{dy}{dx} + 4y^2\right] + \left[x^2\dfrac{dy}{dx} + 2xy\right] = 0$

$2x\dfrac{dy}{dx} - 8xy\dfrac{dy}{dx} + x^2\dfrac{dy}{dx} = -2y + 4y^2 - 2xy$

$(2x - 8xy + x^2)\dfrac{dy}{dx} = -2y + 4y^2 - 2xy \qquad \dfrac{dy}{dx} = \dfrac{-2y + 4y^2 - 2xy}{2x - 8xy + x^2}$

13. $x^{1/2}y^{1/2} = 1$

$x^{1/2} \cdot \dfrac{1}{2}y^{-1/2}\dfrac{dy}{dx} + \dfrac{1}{2}x^{-1/2}y^{1/2} = 0$

$\dfrac{x^{1/2}}{2y^{1/2}}\dfrac{dy}{dx} = \dfrac{-y^{1/2}}{2x^{1/2}}$

$\dfrac{dy}{dx} = \dfrac{-y^{1/2}}{2x^{1/2}} \cdot \dfrac{2y^{1/2}}{x^{1/2}}$

$= \dfrac{-y}{x}$

15. $2(x + y)\left(1 + \dfrac{dy}{dx}\right) = 0$

$1 + \dfrac{dy}{dx} = 0$

$\dfrac{dy}{dx} = -1$

17. $\left[x^2 \cdot 2y\dfrac{dy}{dx} + 2xy^2\right] - 2 = 0$

$2x^2y\dfrac{dy}{dx} = 2 - 2xy^2$

$\dfrac{dy}{dx} = \dfrac{2 - 2xy^2}{2x^2y}$

$= \dfrac{1 - xy^2}{x^2y}$

If $x = 2$: $\qquad 2^2y^2 - 2(2) = 0$

$4y^2 = 4$

$y^2 = 1$

$y = \pm 1$

For $(2,1)$ $\dfrac{dy}{dx}\bigg|_{x=2} = \dfrac{1 - 2(1)^2}{(2)^2(1)}$

$= \dfrac{-1}{4}$

For $(2,-1)$ $\dfrac{dy}{dx}\bigg|_{x=2} = \dfrac{1 - 2(-1)^2}{(2)^2(-1)}$

$= \dfrac{1}{4}$

19a. $18x + 32y\dfrac{dy}{dx} = 0 \qquad 32y\dfrac{dy}{dx} = -18x \qquad \dfrac{dy}{dx} = \dfrac{-9x}{16y}$

19b. If $x = 2$: $\quad 9(2)^2 + 16y^2 = 144 \quad 16y^2 = 108 \quad y^2 = \dfrac{27}{4} \quad y = \dfrac{\pm\sqrt{27}}{2} = \dfrac{\pm 3\sqrt{3}}{2}$

At $\left(2, \dfrac{3\sqrt{3}}{2}\right)$: $\dfrac{dy}{dx} = \dfrac{-9(2)}{16\left(\dfrac{3\sqrt{3}}{2}\right)} = \dfrac{-3}{4\sqrt{3}} = \dfrac{-3\sqrt{3}}{4(3)} = \dfrac{-\sqrt{3}}{4}$

At $\left(2, \dfrac{-3\sqrt{3}}{2}\right)$: $\dfrac{dy}{dx} = \dfrac{-9(2)}{16\left(\dfrac{-3\sqrt{3}}{2}\right)} = \dfrac{\sqrt{3}}{4}$

19c. $\quad y - y_1 = m(x - x_1) \qquad y - \dfrac{3\sqrt{3}}{2} = \dfrac{-\sqrt{3}}{4}(x-2) \qquad 4\left(y - \dfrac{3\sqrt{3}}{2}\right) = -\sqrt{3}(x-2)$

$\qquad 4y - 6\sqrt{3} = -\sqrt{3}x + 2\sqrt{3} \qquad \sqrt{3}x + 4y - 8\sqrt{3} = 0$

19d. $\quad \dfrac{d^2y}{dx^2} = \dfrac{16y(-9) - (-9x)\left(16\dfrac{dy}{dx}\right)}{(16y)^2} = \dfrac{-9y + 9x\dfrac{dy}{dx}}{16y^2} = \dfrac{-9y + 9x\left(\dfrac{-9x}{16y}\right)}{16y^2}$

$\qquad = \dfrac{-144y^2 - 81x^2}{256y^3} = \dfrac{-9(16y^2 + 9x^2)}{256y^3} = \dfrac{-9(144)}{256y^3}$ (from the original equation)

$\qquad = \dfrac{-81}{16y^3}$

21. $\quad 2x\dfrac{dx}{dt} - 18y\dfrac{dy}{dt} = 0$

$\qquad 2(-5)\dfrac{dx}{dt} - 18(1)(10) = 0$

$\qquad -10\dfrac{dx}{dt} = 180$

$\qquad \dfrac{dx}{dt} = -18$

23. $\quad 2x\dfrac{dx}{dt} + 2y\dfrac{dy}{dt} = 0$

$\qquad 2(3)(2) + 2(-4)\dfrac{dy}{dt} = 0$

$\qquad -8\dfrac{dy}{dt} = -12$

$\qquad \dfrac{dy}{dt} = \dfrac{3}{2}$

25. $\quad V = \pi r^2 h$

$\qquad V = \pi(3)^2 h$

$\qquad V = 9\pi h$

$\qquad \dfrac{dy}{dt} = 9\pi\dfrac{dh}{dt}$

$\qquad 3.5 = 9\pi\dfrac{dh}{dt}$

$\qquad \dfrac{dh}{dt} = \dfrac{3.5}{9\pi} \approx 0.124$ ft./hr.

27. $\quad [p(1) + \dfrac{dp}{dx}(x)] + 2 = 0$

$\qquad x\dfrac{dp}{dx} = -p - 2$

$\qquad \dfrac{dp}{dx} = \dfrac{-p-2}{x}$

29. $\quad \dfrac{dP}{dx} = -4x + 140$

$\qquad \dfrac{dP}{dt} = \dfrac{dP}{dx} \cdot \dfrac{dx}{dt}$

$\qquad = (-4x + 140)(3)$

$\qquad = -12x + 420$

$\qquad = -12(20) + 420$

$\qquad = \$180/\text{day}$

31. $\quad \dfrac{dc}{dx} = 2$

$\qquad 2p + 4x\dfrac{dx}{dp} = 0$

$\qquad 4x\dfrac{dx}{dp} = -2p$

$\qquad \dfrac{dx}{dp} = \dfrac{-2p}{4x} = \dfrac{-p}{2x}$

$\qquad \dfrac{dc}{dp} = \dfrac{dc}{dx} \cdot \dfrac{dx}{dp}$

$\qquad = 2\left(\dfrac{-p}{2x}\right) = \dfrac{-p}{x}$

33. $\frac{dc}{dt} = \frac{dw}{dt} + \frac{dA}{dt}$

$\frac{dc}{dt} = 4 + (-1)$

$= 3$

The number of clients is increasing at a rate of 3 clients per hour.

35. $PV = c$

$\left[P\frac{dV}{dt} + \frac{dP}{dt}(V) \right] = 0$

$15(10) + \frac{dP}{dt}(120) = 0$

$120\frac{dP}{dt} = -150$

$\frac{dP}{dt} = \frac{-5}{4}$

$= -1.25 \text{ lb./in.}^2 \text{ per second}$

37. $A = \pi r^2$

$\frac{dA}{dt} = 2\pi r\frac{dr}{dt}$

$= 2\pi(6)(2) = 24\pi \approx 75.4 \text{ in.}^2/\text{min.}$

Section 3.7 (pages 209-211)

1. $f'(x) = -4$

$dy = f'(x)dx$

$= -4dx$

3. $f'(x) = 3x^2 + 1$

$dy = f'(x)dx$

$= (3x^2 + 1)dx$

5. $y = (2 - x)^{-1}$

$f'(x) = -1(2 - x)^{-2}(-1)$

$= \frac{1}{(2 - x)^2}$

$dy = f'(x)dx$

$= \frac{1}{(2 - x)^2}dx$

$= \frac{dx}{(2 - x)^2}$

7. $x = 25$

$dx = 1$

$f(x) = \sqrt{x} = x^{1/2}$

$f'(x) = \frac{1}{2}x^{-1/2} = \frac{1}{2\sqrt{x}}$

$dy = \frac{1}{2\sqrt{x}}dx$

$= \frac{1}{2\sqrt{25}}(1)$

$= 0.1$

$f(26) \approx f(25) + dy$

$= 5 + 0.1 = 5.1$

9. $x = 8$

$dx = 0.1$

$f(x) = \sqrt[3]{x} = x^{1/3}$

$f'(x) = \frac{1}{3}x^{-2/3} = \frac{1}{3\sqrt[3]{x^2}}$

$dy = \frac{1}{3\sqrt[3]{x^2}}dx$

$= \frac{1}{3(\sqrt[3]{8})^2}(0.1)$

≈ 0.00833

$f(8.1) \approx f(8) + dy$

$= 2 + 0.00833$

$= 2.00833$

11. $x = 9$

$dx = 0.1$

$f(x) = \frac{1}{\sqrt{x}} = x^{1/2}$

$f'(x) = \frac{-1}{2}x^{-3/2} = \frac{-1}{2\sqrt{x^3}}$

$dy = \frac{-1}{2\sqrt{x^3}}dx$

$= \frac{-1}{2(\sqrt{9})^3}(0.1) \approx -0.00185$

$f(9.1) \approx f(9) + dy$

$= \frac{1}{3} + (-0.00185)$

≈ 0.33148

13a. $f(x + h) - f(x) = f(2 + 0.1) - f(2)$

$= f(2.1) - f(2) = 9.261 - 8 = 1.261$

13b. $dy = f'(x)dx$

$= 3x^2 dx = 3(2)^2(0.1) = 1.2$

15a. $f(x + h) - f(x) = f[3 + (-0.1)] - f(3)$

$= f(2.9) - f(3) = 42.05 - 45 = -2.95$

15b. $dy = f'(x)dx$

$= 10x\,dx = 10(3)(-0.1) = -3$

17a. $f(x + h) - f(x) = f(4 + 0.01) - f(4)$

$= f(4.01) - f(4) \approx 1.99750156 - 2 = -0.00249844$

17b. $dy = f'(x)dx$

$= -\frac{1}{2}x^{-1/2}dx = \frac{-1}{2\sqrt{x}}dx = \frac{-1}{2\sqrt{4}}(0.01) = -0.0025$

19a. $f(x + h) - f(x) = f[2 + (-0.01)] - f(2)$

$= f(1.99) - f(2) \approx 0.252518875 - 0.25 = 0.002518876$

19b. $dy = f'(x)dx$

$= -2x^{-3}dx = \frac{-2}{x^3}dx = \frac{-2}{(2)^3}(-0.01) = 0.0025$

21. $\dfrac{dy}{dx} = 1 - 2x$

$dy = (1 - 2x)dx$

23. $y = (x - 3)^{1/2}$

$\dfrac{dy}{dx} = \frac{1}{2}(x - 3)^{-1/2}$

$dy = \dfrac{1}{2\sqrt{x - 3}}dx$

25. $x = 5$

$dx = 0.1$

$V = x^3$

$dV = 3x^2 dx$

$= 3(5)^2(0.1) = 7.5 \text{ cm}^3$

27. $r = 6$

$dr = -0.2$

$A = \pi r^2$

$dA = 2\pi r dr$

$= 2\pi(6)(-0.2) \approx -7.54 \text{ in.}^2$

29. $r = \ = 3$

$dr = 0.01$

$V = \frac{4}{3}\pi r^3$

$dV = 4\pi r^2 dr$

$= 4\pi(3)^2(0.01)$

$\approx 1.131 \text{ cm}^3$

31. $r_2 = 6$

$dr_2 = 0.2$

$s = \pi(r_2^2 - r_1^2)$

$= \pi(r_2^2 - 2^2) = \pi r_2^2 - 4\pi$

$dS = 2\pi r_2 dr_2$

$= 2\pi(6)(0.2) \approx 7.54 \text{ cm}^2$

33. $x = 81$

$dx = -1$

$f(x) = 50 - \sqrt{x}$

$= 50 - x^{1/2}$

$f'(x) = \dfrac{-1}{2}x^{-1/2}$

$= \dfrac{-1}{2\sqrt{x}}$

$dy = \dfrac{-1}{2\sqrt{x}}dx$

$= \dfrac{-1}{2\sqrt{81}}(-1) \approx 0.06$

$f(80) \approx f(81) + dy = \$41.06$

35. $r = \frac{4}{2} = 2$

$dr = \dfrac{-0.2}{2} = -0.1$

$A = \pi r^2$

$dA = 2\pi r dr$

$= 2\pi(2)(-0.1)$

$\approx -1.26 \text{ mm}^2$

37. $w = 4, \ dw = 0.25$

$A = lw = 12w$

$dA = 12dw$

$= 12(0.25) = 3 \text{ in.}^2$

Chapter 3 Review   (pages 212-214)

1. $\dfrac{dy}{dx} = 10x - 7$

2. $\dfrac{dy}{dx} = -3x^2$

3. $$\frac{dy}{dx} = \frac{(x^2+2)(1)-(x+1)(2x)}{(x^2+2)^2} = \frac{x^2+2-2x^2-2x}{(x^2+2)^2} = \frac{-x^2-2x+2}{(x^2+2)^2}$$

4. $$y = x^2(1-x^2)^{1/2} \qquad u(x) = x^2 \qquad v(x) = (1-x^2)^{1/2}$$

$$\begin{aligned}
\frac{dy}{dx} &= x^2\left[\frac{1}{2}(1-x^2)^{-1/2}(-2x)\right]+2x(1-x^2)^{1/2}\\
&= x(1-x^2)^{-1/2}[x^2(-1)+2(1-x^2)]\\
&= x(1-x^2)^{-1/2}(-x^2+2-2x^2)\\
&= \frac{x(-3x^2+2)}{\sqrt{1-x^2}}
\end{aligned}$$

5. $$y = x^2 - x^{-1/2}$$

$$\frac{dy}{dx} = 2x - (-\tfrac{1}{2})x^{-3/2}$$

$$= 2x + \frac{1}{2\sqrt{x^3}}$$

6. $$2x - 6y\frac{dy}{dx} = 5$$

$$-6y\frac{dy}{dx} = 5 - 2x$$

$$\frac{dy}{dx} = \frac{5-2x}{-6y} \quad \text{or} \quad \frac{2x-5}{6y}$$

7a. $$\frac{dy}{dx} = \frac{(x-3)(12)-12x(1)}{(x-3)^2}$$

$$= \frac{-36}{(x-3)^2}$$

$$= \frac{dy}{dx}\bigg|_{x=-1} = \frac{-36}{16} = \frac{-9}{4}$$

7b. $$y - 3 = \frac{-9}{4}[x-(-1)]$$

$$4(y-3) = -9(x+1)$$

$$4y - 12 = -9x - 9$$

$$9x + 4y - 3 = 0$$

8. $$f'(x) = -4x + 11 \qquad f'(0) = -4(0)+11 = 11$$

9a.
$$y = (u+2)^{-1} \qquad u = x^{1/2} \qquad \frac{dy}{dx} = \frac{dy}{du}\cdot\frac{du}{dx}$$

$$\frac{dy}{du} = -1(u+2)^{-2} \qquad \frac{du}{dx} = \frac{1}{2}x^{-1/2} \qquad = \frac{-1}{(u+2)^2}\cdot\frac{1}{2\sqrt{x}}$$

$$= \frac{-1}{(u+2)^2} \qquad\qquad = \frac{1}{2\sqrt{x}} \qquad\qquad = \frac{-1}{(\sqrt{x}+2)^2(2\sqrt{x})}$$

9b. $$y = \frac{1}{\sqrt{x}+2} = \frac{1}{x^{1/2}+2} = (x^{1/2}+2)^{-1}$$

$$\frac{dy}{dx} = -1(x^{1/2}+2)^{-2}\left(\tfrac{1}{2}x^{-1/2}\right) = \frac{-1}{(\sqrt{x}+2)(2\sqrt{x})}$$

10. $f'(x) = \dfrac{(x^2-1)(1) - x(2x)}{(x^2-1)^2} = \dfrac{-1-x^2}{(x^2-1)^2}$

$f''(x) = \dfrac{(x^2-1)^2(-2x) - (-1-x^2)[2(x^2-1)(2x)]}{\left[(x^2-1)^2\right]^2}$

$\qquad = \dfrac{-2x(x^2-1)[(x^2-1) + (-1-x^2)(2)]}{(x^2-1)^4} = \dfrac{-2x(x^2-1)(x^2-1-2-2x^2)}{(x^2-1)^4}$

$\qquad = \dfrac{-2x(-3-x^2)}{(x^2-1)^3} = \dfrac{2x(x^2+3)}{(x^2-1)^3}$

11. $\dfrac{dy}{dx} = 4(2x+1)^3(2) = 8(2x+1)^3$

$\dfrac{d^2y}{dx^2} = 8(3)(2x+1)^2(2) = 48(2x+1)^2$

$\dfrac{d^3y}{dx^3} = 48(2)(2x+1)(2) = 192(2x+1)$

$\left. \dfrac{d^3y}{dx^3}\right|_{x=0} = 192[2(0)+1] = 192$

12a. $C(50) = 100 + 25(50) + 0.1(50)^2 = \$1600$

12b. $C'(x) = 25 + 0.2x$ $\qquad\qquad$ 12c. $\quad C'(50) = 25 + 0.2(50) = \$35$

13a. $R(x) = px = (16 - 0.02x)(x) = 16x - 0.02x^2$

13b. $R'(x) = 16 - 0.04x$

13c. $P(x) = R(x) - C(x)$

$\qquad = (16x - 0.02x^2) - (0.004x^2 + 1000) = -0.024x^2 + 16x - 1000$

13d. $P'(x) = -0.048x + 16$

14. $P(x) = 700(x^2 + 16)^{-1/2}$

$P'(x) = 700(\tfrac{-1}{2})(x^2+16)^{-3/2}(2x) = \dfrac{-700x}{\sqrt{(x^2+16)^3}}$

$P'(3) = \dfrac{-700(3)}{\left(\sqrt{(3)^2+16}\right)^3} = -\$16.80$

15. $N'(t) = -1 + 0.05t$ $\qquad\qquad$ 16. $\qquad V = \tfrac{4}{3}\pi r^3$

$\quad N'(10) = -1 + 0.05(10) = -0.5$ $\qquad\qquad\quad dV = 4\pi r^2 dr$

$\qquad\qquad\qquad\qquad\qquad\qquad\qquad\qquad 20 = 4\pi(5)^2 dr$

$\qquad\qquad\qquad\qquad\qquad\qquad\qquad\quad dr = \dfrac{20}{100\pi} = \dfrac{1}{5\pi} \text{ ft./sec.}$

17a. $\dfrac{dh}{dt} = -32t$ $\qquad \dfrac{dh}{dt}\Big|_{t=0.5} = -32(0.5)$ $\qquad = -16$ ft./sec.

17b. $0 = 32 - 16t^2$

$16t^2 = 32,\ t^2 = 2,\ t = \sqrt{2} \approx 1.4$ sec.

18. $\qquad x = 4$

$\qquad dx = 0.1$

$\qquad f(x) = \sqrt{x} = x^{1/2}$

$\qquad dy = \frac{1}{2}x^{-1/2}dx = \dfrac{1}{2\sqrt{x}}dx = \dfrac{1}{2\sqrt{4}}(0.1) = 0.025$

$\qquad f(4.1) \approx f(4) + dy$

$\qquad\qquad = 2 + 0.025 + 2.025$

Chapter 3 Test $\quad$ (pages 214-216)

1. $\qquad y = 3x^2 + 2x^{-2}$

$\qquad \dfrac{dy}{dx} = 6x - 4x^{-3}$

$\qquad\qquad = 6x - \dfrac{4}{x^3} \qquad$ (a)

2. $\qquad f(x) = (2x+5)^{1/2}$

$\qquad f'(x) = \frac{1}{2}(2x+5)^{-1/2}(2)$

$\qquad\qquad = \dfrac{1}{\sqrt{2x+5}}$

$\qquad f'(2) = \dfrac{1}{\sqrt{2(2)+5}} = \frac{1}{3} \qquad$ (d)

3. $\qquad u(x) = 2x^3 \qquad v(x) = (1-2x)^5$

$\qquad f'(x) = 2x^3[5(1-2x)^4(-2)] + 6x^2(1-2x)^5$

$\qquad\qquad = 2x^2(1-2x)^4[x(5)(-2) + 3(1-2x)]$

$\qquad\qquad = 2x^2(1-2x)^4(-10x + 3 - 6x)$

$\qquad\qquad = 2x^2(1-2x)^4(3-16x)$

$\qquad f'(1) = 2(1)^2[1-2(1)]^4[3-16(1)]$

$\qquad\qquad = 2(1)(-1)^4(-13) = -26 \qquad$ (c)

4. $\qquad f'(x) = 4(3)[g(x)]^2 g'(x)$

$\qquad\qquad = 12[g(x)]^2 g'(x) \qquad$ (b)

5. $\qquad 3x^2 + 3y^2\dfrac{dy}{dx} = 2\dfrac{dy}{dx}$

$\qquad\qquad 3x^2 = 2\dfrac{dy}{dx} - 3y^2\dfrac{dy}{dx}$

$\qquad\qquad 3x^2 = (2-3y^2)\dfrac{dy}{dx}$

$\qquad\qquad \dfrac{3x^2}{2-3y^2} = \dfrac{dy}{dx} \qquad$ (b)

6. $\qquad \dfrac{dy}{dx} = \dfrac{(1-x)(2x) - x^2(-1)}{(1-x)^2} = \dfrac{2x - 2x^2 + x^2}{(1-x)^2} = \dfrac{2x - x^2}{(1-x)^2} \qquad$ (a)

7. $f'(x) = 4(x^3 + 2)^3(3x^2) = 12x^2(x^3 + 2)^3$

$f''(x) = 12x^2[3(x^3 + 2)^2(3x^2)] + 24x(x^3 + 2)^3$

$\qquad = 12x(x^2 + 2)^2[x(3)(3x^2) + 2(x^3 + 2)]$

$\qquad = 12x(x^2 + 2)^2(9x^3 + 2x^3 + 4)$

$\qquad = 12x(x^2 + 2)^2(11x^3 + 4) \qquad$ (d)

8. $y = 2(x - 1)^{-1}$

$\dfrac{dy}{dx} = 2(-1)(x - 1)^{-2} = \dfrac{-2}{(x - 1)^2}$

$\dfrac{dy}{dx}\bigg|_{x=2} = \dfrac{-2}{(2 - 1)^2} = -2$

$y - 2 = -2(x - 2)$

$y - 2 = -2x + 4$

$\qquad y = -2x + 6 \qquad$ (b)

9. $f'(x) = \dfrac{x^2(1) - (x - 1)(2x)}{(x^2)^2}$

$\qquad = \dfrac{x^2 - 2x^2 + 2x}{x^4}$

$\qquad = \dfrac{2x - x^2}{x^4} = \dfrac{2 - x}{x^3}$

$2 - x = 0$

$x = 2 \qquad$ (b)

10. $\dfrac{dy}{du} = 3u^2 + 1$

$u = 2x^{1/2}$

$\dfrac{du}{dx} = 2(\tfrac{1}{2})x^{-1/2}$

$\qquad = \dfrac{1}{\sqrt{x}}$

$\dfrac{dy}{dx} = \dfrac{dy}{du} \cdot \dfrac{du}{dx}$

$\qquad = (3u^2 + 1)\left(\dfrac{1}{\sqrt{x}}\right)$

$\qquad = [3(2\sqrt{x})^2 + 1]\left(\dfrac{1}{\sqrt{x}}\right)$

$\qquad = \dfrac{12x + 1}{\sqrt{x}} \qquad$ (a)

11. $f(x) = (1 - 2x)^{-1/2}$

$f'(x) = \dfrac{-1}{2}(1 - 2x)^{-3/2}(-2)$

$\qquad = \dfrac{1}{\sqrt{(1 - 2x)^3}}$

$f'(-4) = \dfrac{1}{(\sqrt{1 - 2(-4)})^3} = \dfrac{1}{27}$

12. $f(x) = x - x^{-2}$

$f'(x) = 1 - (-2)x^{-3}$

$\qquad = 1 + \dfrac{2}{x^3}$

13. $f'(x) = 3x^2 - 3$

$3x^2 - 3 = 0,\ 3x^2 = 3,\ x^2 = 1,\ x = \pm 1$

14. $f'(x) = 4(1 - x^3)^3(-3x^2) = -12x^2(1 - x^3)^3$

$f''(x) = -12x^2[3(1 - x^3)^2(-3x^2)] + (-24x)(1 - x^3)^3$

$\qquad = -12x(1 - x^3)^2[x(3)(-3x^2) + 2(1 - x^3)]$

$\qquad = -12x(1 - x^3)^2(-9x^3 + 2 - 2x^3) = -12x(1 - x^3)^2(-11x^3 + 2)$

$f''(0) = 0$

15a. $\quad y = 3x^{-1/3}$

$\qquad \dfrac{dy}{dx} = 3\left(\dfrac{-1}{3}\right)x^{-4/3}$

$\qquad = -1x^{-4/3} = \dfrac{-1}{\sqrt[3]{x^4}}$

15b. $\quad \dfrac{d^2y}{dx^2} = (-1)\left(\dfrac{-4}{3}\right)x^{-7/3}$

$\qquad = \dfrac{4}{3}x^{-7/3} = \dfrac{4}{3\sqrt[3]{x^7}}$

15c. $\quad \dfrac{d^3y}{dx^3} = \dfrac{4}{3}\left(\dfrac{-7}{3}\right)x^{-10/3} = \dfrac{-28}{9}x^{-10/3} = \dfrac{-28}{9\sqrt[3]{x^{10}}}$

16. $\quad C'(x) = 0.15x^2$

$\qquad C'(5) = 0.15(5)^2$

$\qquad = 3.75$

$\qquad (\$3750)$

17a. $\quad P(x) = 3(10x - 100)^{1/2}$

$\qquad P'(x) = 3(\tfrac{1}{2})(10x - 100)^{-1/2}(10)$

$\qquad = \dfrac{15}{\sqrt{10x - 100}}$

17b. $\quad P'(100) = \dfrac{15}{\sqrt{10(100) - 100}} = \dfrac{1}{2}$

18. $\quad Q(t) = 40 - \dfrac{1}{120}t^2$

$\qquad Q'(t) = \dfrac{-1}{120}(2t)$

$\qquad = \dfrac{-t}{60}$

$\qquad Q'(30) = \dfrac{-30}{60}$

$\qquad = -0.5 \text{ milligram/min.}$

19. $\quad N(t) = 12(35 - 2t)^{1/2}$

$\qquad N' = 12\left(\tfrac{1}{2}\right)(35 - 2t)^{-1/2}(-2)$

$\qquad = \dfrac{-12}{\sqrt{35 - 2t}}$

$\qquad N'(5) = \dfrac{-12}{\sqrt{35 - 2(5)}}$

$\qquad = -2.4 \quad (-240)$

20a. $\quad v(t) = s'(t)$

$\qquad = 4t^3 + 2t$

$\qquad v(2) = 4(2)^3 + 2(2)$

$\qquad = 36 \text{ ft./sec.}$

20b. $\quad a(t) = v'(t)$

$\qquad = 12t^2 + 2$

$\qquad a(2) = 12(2)^2 + 2$

$\qquad = 50 \text{ ft./sec.}^2$

CHAPTER 4: CURVE SKETCHING AND OPTIMIZATION

Section 4.1 (pages 226-229)

1. increasing $(-\infty, 2)$ 3.   increasing for all x
    decreasing $(2, \infty)$

5. $\dfrac{dy}{dx} = 3$ increasing for all x

7. $f'(x) = -2x$

    Set $f'(x) = 0$: $-2x = 0,\ x = 0$

interval	$x < 0$	$x > 0$
sign of $f'(x)$	+	−
result: $f(x)$ is	increasing	decreasing

9. $f'(x) = 3x^2$

    Set $f'(x) = 0$: $3x^2 = 0,\ x = 0$

interval	$x < 0$	$x > 0$
sign of $f'(x)$	+	+
result: $f(x)$ is	increasing	increasing

11. $\dfrac{dy}{dx} = 5x^4 - 5$

    Set $\dfrac{dy}{dx} = 0$: $5x^4 - 5 = 0$
    $$5(x^4 - 1) = 0$$
    $$5(x^2 + 1)(x^2 - 1) = 0$$
    $$5(x^2 + 1)(x + 1)(x - 1) = 0$$
    $$x + 1 = 0,\ x = -1 \qquad x - 1 = 0,\ x = 1$$

interval	$x < -1$	$-1 < x < 1$	$x > 1$
sign of $f'(x)$	+	−	+
result: $f(x)$ is	increasing	decreasing	increasing

13. $f'(x) = \dfrac{(x-4)(2) - 2x(1)}{(x-4)^2} = \dfrac{-8}{(x-4)^2}$

    $f'(x)$ is undefined for $x = 4$, but $x = 4$ is not a critical value since $f(4)$ is undefined.

interval	$x < 4$	$x > 4$
sign of $f'(x)$	−	−
result: $f(x)$ is	decreasing	decreasing

15. $y = (x+2)^{-1/2}$ $\dfrac{dy}{dx} = \dfrac{-1}{2}(x+2)^{-3/2} = \dfrac{-1}{\sqrt{(x+2)^3}}$

$\dfrac{dy}{dx}$ is never 0. It is undefined for $x \le -2$.

interval	$x > -2$
sign of $f'(x)$	$-$
result: $f(x)$ is	decreasing

17. $f'(x) = 3(x-2)^2$

Set $f'(x) = 0$: $3(x-2)^2 = 0$, $x = 2$

interval	$x < 2$	$x > 2$
sign of $f'(x)$	$+$	$+$
result: $f(x)$ is	increasing	increasing

x	y
-1	-27
0	-8
1	-1
2	0
3	1
4	8

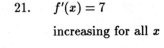

19. $f'(x) = -2x$

Set $f'(x) = 0$: $-2x = 0$, $x = 0$

interval	$x < 0$	$x > 0$
sign of $f'(x)$	$+$	$-$
result: $f(x)$ is	increasing	decreasing

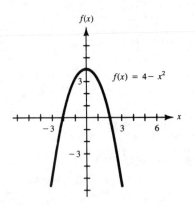

21. $f'(x) = 7$

increasing for all x

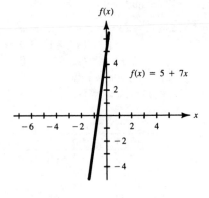

23. $f'(x) = 3x^2 - 12x + 11$

Set $f'(x) = 0$: $3x^2 - 12x + 11 = 0$, $x = \dfrac{-(-12) \pm \sqrt{(-12)^2 - 4(3)(11)}}{2(3)}$

$$= \frac{12 \pm \sqrt{12}}{6} = \frac{12 \pm 2\sqrt{3}}{6} = \frac{6 \pm \sqrt{3}}{3}$$

interval	$x < \dfrac{6-\sqrt{3}}{3}$	$\dfrac{6-\sqrt{3}}{3} < x < \dfrac{6+\sqrt{3}}{3}$	$x > \dfrac{6+\sqrt{3}}{3}$
sign of $f'(x)$	$+$	$-$	$+$
result: $f(x)$ is	increasing	decreasing	increasing

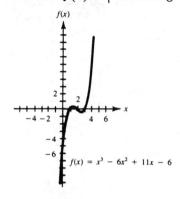

$f(x) = x^3 - 6x^2 + 11x - 6$

25. $f'(x) = 3x^2 - 3$

Set $f'(x) = 0$ $3x^2 - 3 = 0$, $3(x^2 - 1) = 0$, $3(x+1)(x-1) = 0$

$$x = -1 \qquad x = 1$$

interval	$x < -1$	$-1 < x < 1$	$x > 1$
sign of $f'(x)$	$+$	$-$	$+$
result: $f(x)$ is	increasing	decreasing	increasing

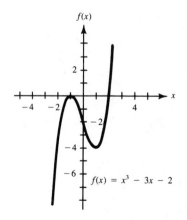

$f(x) = x^3 - 3x - 2$

27.	$f'(x) = (x-2)^2(-1) + 2(x-2)(1-x)$

　　　$= (x-2)[(x-2)(-1) + 2(1-x)]$

　　　$= (x-2)(4-3x)$

Set $f'(x) = 0$:

$(x-2)(4-3x) = 0$

$x = 2 \qquad x = \dfrac{4}{3}$

interval	$x < \frac{4}{3}$	$\frac{4}{3} < x < 2$	$x > 2$
sign of $f'(x)$	−	+	−
result: $f(x)$ is	decreasing	increasing	decreasing

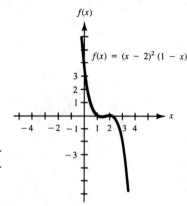

$f(x)$

$f(x) = (x-2)^2(1-x)$

29a.	$f'(x) = \dfrac{(3-x)(0) - 2(-1)}{(3-x)^2} = \dfrac{2}{(3-x)^2}$ 　　　29c.

29b.	No, since $f(3)$ does not exist.

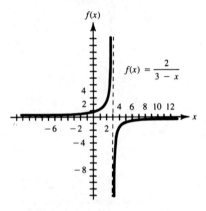

$f(x)$

$f(x) = \dfrac{2}{3-x}$

31.	$x = 6 \qquad \$30,000$

33a.	$C'(x) = 0.8x - 1.6$

33b.	Set $C'(x) = 0$: 　　$0.8x - 1.6 = 0$, $x = 2$　(Recall that $x > 0$)

interval	$0 < x < 2$	$x > 2$
sign of $f'(x)$	−	+
result: $f(x)$ is	decreasing	increasing

35a.	$P(x) = R(x) - C(x) = (700x - 0.2x^2) - (0.3x^2 + 200x + 500) = -0.5x^2 + 500x - 500$

35b.	$P'(x) = -x + 500$

Set $P'(x) = 0$: 　　$-x + 500 = 0$, $x = 500$ (Recall that $x > 0$)

interval	$0 < x < 500$	$x > 500$
sign of $f'(x)$	+	−
result: $f(x)$ is	increasing	decreasing

37. $N'(t) = \dfrac{(t^2+4)(200)-(200t)(2t)}{(t^2+4)^2}+0=\dfrac{800-200t^2}{(t^2+4)^2}$

Set $N'(t)=0$: $\quad \dfrac{800-200t^2}{(t^2+4)^2}=0,\ 800-200t^2=0,\ 200(4-t^2)=0$

$$t=2 \text{ since } 0 \le t \le 15$$

interval	$0<t<2$	$2<t<15$
sign of $f'(x)$	$+$	$-$
result: $f(x)$ is	increasing	decreasing

(increasing between 1970 and 1972)

Section 4.1A   (page 235)

1. $\dfrac{dy}{dx}=-2x$

Set $\dfrac{dy}{dx}=0$: $\quad -2x=0$

$$x=0$$

increasing: $x<0$

decreasing: $x>0$

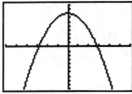

3. $\dfrac{dy}{dx}=4x^3+4x^2-8x$

Set $\dfrac{dy}{dx}=0$: $\quad 4x^3+4x^2-8x=0$

$$4x(x^2+x-2)=0$$

$$4x(x+2)(x-1)=0$$

$$x=0,\ x=-2,\ x=1$$

increasing: $-2<x<0,\ x>1$

decreasing: $x<-2,\ 0<x<1$

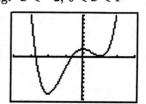

5. $f'(x)=(x^2-2)(2x)+(2x)(x^2-1)$

$\qquad = 2x[(x^2-2)+(x^2-1)]$

$\qquad = 2x(2x^2-3)$

Set $f'(x)=0$: $\quad 2x(2x^2-3)=0$

$x=0,\ x=\pm\sqrt{\dfrac{3}{2}}=\pm\sqrt{1.5}$

increasing: $-\sqrt{1.5}<x<0$

$$x>\sqrt{1.5}$$

decreasing: $x<-\sqrt{1.5}$

$$0<x<\sqrt{1.5}$$

7. $y = (x-3)^{1/3}$

$\frac{dy}{dx} = \frac{1}{3}(x-3)^{-2/3} = \frac{1}{3\sqrt[3]{(x-3)^2}}$

$\frac{dy}{dx}$ is not defined at $x = 3$.

increasing everywhere

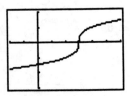

9. $f(x) = (x+1)^{-1/2}$

$f'(x) = \frac{-1}{2}(x+1)^{-3/2}$

$\quad = \frac{-1}{2\sqrt{(x+1)^3}}$

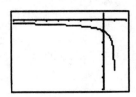

$f'(x)$ is not defined for $x \leq -1$.

decreasing: $x > -1$

11a. $f'(x) = \frac{(x-4)(0) - 3(1)}{(x-4)^2} = \frac{-3}{(x-4)^2}$

11b. No, since $f(4)$ is not defined.

11c.

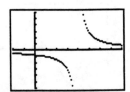

13. function: g

derivative: f

15. function: f

derivative: g

Section 4.2 (pages 239-241)

1. $\frac{dy}{dx} = 2x$

$\frac{d^2y}{dx^2} = 2$

concave up for all x

3. $\frac{dy}{dx} = 2 - 2x$

$\frac{d^2y}{dx^2} = -2$

concave down for all x

5. $\frac{dy}{dx} = 2x - 4$

$\frac{d^2y}{dx^2} = 2$

concave up for all x

7. $\frac{dy}{dx} = 3x^2 + 6x \qquad \frac{d^2y}{dx^2} = 6x + 6$

Set $\frac{d^2y}{dx^2} = 0$: $\quad 6x + 6 = 0, \; x = -1$

interval	$x < -1$	$x > -1$
sign of $\frac{d^2y}{dx^2}$	$-$	$+$
result: y is	concave down	concave up

9. $\dfrac{dy}{dx} = 4x^3 + 4$

$\dfrac{d^2y}{dx^2} = 12x^2$

Set $\dfrac{d^2y}{dx^2} = 0$: $12x^2 = 0$, $x = 0$

interval	$x < 0$	$x > 0$
sign of $\dfrac{d^2y}{dx^2}$	$+$	$+$
result: y is	concave up	concave up

11. $f(x) = x^2 + x^{-1}$

$f'(x) = 2x - x^{-2}$

$f''(x) = 2 + 2x^{-3} = 2 + \dfrac{2}{x^3}$

Set $f''(x) = 0$: $\quad 2 + \dfrac{2}{x^3} = 0$, $\dfrac{2x^3 + 2}{x^3} = 0$, $2x^3 + 2 = 0$, $2(x^3 + 1) = 0$

$\qquad x^3 = -1$, $x = -1$ $\quad$ $f(x)$ is not defined at $x = 0$

interval	$x < -1$	$-1 < x < 0$	$x > 0$
sign of $f''(x)$	$+$	$-$	$+$
result: $f(x)$ is	concave up	concave down	concave up

13. $\dfrac{dy}{dx} = -4x - 6x^2$

$\dfrac{d^2y}{dx^2} = -4 - 12x$

Set $\dfrac{d^2y}{dx^2} = 0$: $\quad -4 - 12x = 0$, $x = \dfrac{-1}{3}$

interval	$x < \frac{-1}{3}$	$x > \frac{-1}{3}$
sign of $\dfrac{d^2y}{dx^2}$	$+$	$-$
result: y is	concave up	concave down

point of inflection $\left(\dfrac{-1}{3}, \dfrac{131}{27}\right)$

15. $y = x^{1/3}$

$\dfrac{dy}{dx} = \dfrac{1}{3}x^{-2/3}$

$\dfrac{d^2y}{dx^2} = \dfrac{1}{3}\left(\dfrac{-2}{3}\right)x^{-5/3} = \dfrac{-2}{9\sqrt[3]{x^5}}$

$\dfrac{d^2y}{dx^2}$ is undefined for $x = 0$

interval	$x < 0$	$x > 0$
sign of $\dfrac{d^2y}{dx^2}$	$+$	$-$
result: y is	concave up	concave down

point of inflection $(0,0)$

17. $\dfrac{dy}{dx} = 4 - 3x^2$

$\dfrac{d^2y}{dx^2} = 6x$

Set $\dfrac{d^2y}{dx^2} = 0$: $\quad -6x = 0$, $x = 0$

interval	$x < 0$	$x > 0$
sign of $\dfrac{d^2y}{dx^2}$	$+$	$-$
result: y is	concave up	concave down

point of inflection $(0,0)$

19. $\dfrac{dy}{dx} = 3x^2 + 14x + 11$

$\dfrac{d^2y}{dx^2} = 6x + 14$

Set $\dfrac{d^2y}{dx^2} = 0$: $6x + 14 = 0$, $x = \dfrac{-7}{3}$

interval	$x < \frac{-7}{3}$	$x > \frac{-7}{3}$
sign of $\frac{d^2y}{dx^2}$	$-$	$+$
result: y is	concave down	concave up

point of inflection $\left(\dfrac{-7}{3}, \dfrac{128}{3}\right)$

21a. $f'(x) = 2(x+1) = 2x + 2$

Set $f'(x) = 0$: $2(x+1) = 0$

$x = -1$

interval	$x < -1$	$x > -1$
sign of $f'(x)$	$-$	$+$
result: $f(x)$ is	decreasing	increasing

21b. $f''(x) = 2$

concave up for all x

21c. none

23a. $f'(x) = x^2 + 2x - 8$

Set $f'(x) = 0$: $x^2 + 2x - 8 = 0$, $(x+4)(x-2) = 0$

$x = -4$, $x = 2$

interval	$x < -4$	$-4 < x < 2$	$x > 2$
sign of $f'(x)$	$+$	$-$	$+$
result: $f(x)$ is	increasing	decreasing	increasing

23b. $f''(x) = 2x + 2$

Set $f''(x) = 0$: $2x + 2 = 0$

$x = -1$

interval	$x < -1$	$x > -1$
sign of $f''(x)$	$-$	$+$
result: $f(x)$ is	concave down	concave up

23c. point of inflection $\left(-1, \dfrac{29}{3}\right)$

25a. $\dfrac{dy}{dx} = \dfrac{2}{3}x^{-1/3} = \dfrac{2}{3\sqrt[3]{x}}$

$\dfrac{dy}{dx}$ is undefined for $x = 0$

interval	$x < 0$	$x > 0$
sign of $\frac{dy}{dx}$	$-$	$+$
result: y is	decreasing	increasing

25b. $\dfrac{d^2y}{dx^2} = \dfrac{2}{3}\left(\dfrac{-1}{3}\right)x^{-4/3} = \dfrac{-2}{9\sqrt[3]{x^4}}$

$\dfrac{d^2y}{dx^2}$ is undefined for $x = 0$

interval	$x < 0$	$x > 0$
sign of $\frac{d^2y}{dx^2}$	$-$	$-$
result: y is	concave down	concave down

25c. none

27. $f'(x) = m$, $f''(x) = 0$
neither concave up nor concave down

29.

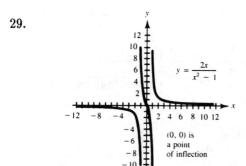

$$\frac{dy}{dx} = \frac{(x^2-1)(2) - 2x(2x)}{(x^2-1)^2} = \frac{-2x^2-2}{(x^2-1)^2} = \frac{-2(x^2+1)}{(x^2-1)^2}$$

y is undefined at $x = \pm 1$

interval	$x < -1$	$-1 < x < 1$	$x > 1$
sign of $\dfrac{dy}{dx}$	$-$	$-$	$-$
result: y is	decreasing	decreasing	decreasing

$$\frac{d^2y}{dx^2} = \frac{(x^2-1)^2(-4x) - (-2)(x^2+1)[(2)(x^2-1)(2x)]}{[(x^2-1)^2]^2}$$

$$= \frac{-4x(x^2-1)[(x^2-1) - (x^2+1)(2)]}{(x^2-1)^4} = \frac{-4x(-x^2-3)}{(x^2-1)^3} \quad \text{or} \quad \frac{4x(x^2+3)}{(x^2-1)^3}$$

Set $\dfrac{d^2y}{dx^2} = 0$: $\quad \dfrac{4x(x^2+3)}{(x^2-1)^3} = 0$, $4x(x^2+3) = 0$, $4x = 0$, $x = 0$

y is undefined at $x = \pm 1$

interval	$x < -1$	$-1 < x < 0$	$0 < x < 1$	$x > 1$
sign of $\dfrac{d^2y}{dx^2}$	$-$	$+$	$-$	$+$
result: y is	concave down	concave up	concave down	concave up

point of inflection $(0,0)$

31.
$$C(x) = 120 + x^{1/2}$$
$$C'(x) = \frac{1}{2}x^{-1/2}$$
$$C''(x) = \frac{1}{2}\left(\frac{-1}{2}\right)x^{-3/2} = \frac{-1}{4\sqrt{x^3}}$$

This is negative for $x > 0$ so $C(x)$ is concave down.

33a,b.
$$P'(x) = 12x - 9x^2$$
$$P''(x) = 12 - 18x$$
Set $P''(x) = 0$: $12 - 18x = 0$
$$x = \frac{2}{3}$$

interval	$0 < x < \frac{2}{3}$	$x > \frac{2}{3}$
sign of $P''(x)$	$+$	$-$
result: $P'(x)$ is	increasing	decreasing

Section 4.3   (pages 249-253)

1.
$f'(x) = 2x - 2$
Set $f'(x) = 0$: $2x - 2 = 0$
$x = 1$
(critical value)

interval	$x < 1$	$x > 1$
sign of $f'(x)$	$-$	$+$
location	left of 1	right of 1

local minimum $(1, -1)$

3.
$f'(x) = 3x^2 - 4x$

Set $f'(x) = 0$: $3x^2 - 4x = 0$, $x(3x - 4) = 0$, $x = 0$, $x = \frac{4}{3}$ (critical values)

interval	$x < 0$	$0 < x < \frac{4}{3}$	$x > \frac{4}{3}$
sign of $f'(x)$	$+$	$-$	$+$
location	left of 0	right of 0 left of $\frac{4}{3}$	right of $\frac{4}{3}$

local maximum $(0, 0)$

local minimim $\left(\frac{4}{3}, \frac{-32}{27}\right)$

5.
$f'(x) = 12x^3 - 24x^2$

Set $f'(x) = 0$: $12x^3 - 24x^2 = 0$, $12x^2(x - 2) = 0$

$x = 0$, $x = 2$ (critical values)

interval	$x < 0$	$0 < x < 2$	$x > 2$
sign of $f'(x)$	$-$	$-$	$+$
location	left of 0	right of 0 left of 2	right of 2

local minimum $(2, -14)$

7. $f'(x) = 4x^3 - 12x$

Set $f'(x) = 0$: $4x^3 - 12x = 0,\ 4x(x^2 - 3) = 0$

$x = 0,\ x = \pm\sqrt{3}$ (critical values)

interval	$x < -\sqrt{3}$	$-\sqrt{3} < x < 0$	$0 < x < \sqrt{3}$	$x > \sqrt{3}$
sign of $f'(x)$	$-$	$+$	$-$	$+$
location	left of $-\sqrt{3}$	right of $-\sqrt{3}$ left of 0	right of 0 left of $\sqrt{3}$	right of $\sqrt{3}$

local minima $(-\sqrt{3}, -7)\ (\sqrt{3}, -7)$

local maximim $(0, 2)$

9. $f'(x) = \frac{2}{3}(x+3)^{-1/3} = \frac{2}{3\sqrt[3]{x+3}}$

$f'(-3)$ is not defined.

$(x = -3$ critical value$)$

interval	$x < -3$	$x > -3$
sign of $f'(x)$	$-$	$+$
location	left of -3	right of -3

local minimum $(-3, 0)$

11. $\frac{dy}{dx} = \frac{(x^2+1)(-1) - (-x)(2x)}{(x^2+1)^2} = \frac{x^2 - 1}{(x^2+1)^2}$

Set $\frac{dy}{dx} = 0$: $\frac{x^2 - 1}{(x^2+1)^2} = 0,\ x^2 - 1 = 0,\ x = \pm 1$ (critical values)

interval	$x < -1$	$-1 < x < 1$	$x > 1$
sign of $f'(x)$	$+$	$-$	$+$
location	left of -1	right of -1 left of 1	right of 1

local maximum $(-1, \frac{1}{2})$

local minimum $(1, \frac{-1}{2})$

13. $f'(x) = \frac{(x-1)(2x) - x^2(1)}{(x-1)^2} = \frac{x^2 - 2x}{(x-1)^2}$

Set $f'(x) = 0$: $\frac{x^2 - 2x}{(x-1)^2} = 0,\ x^2 - 2x = 0,\ x(x-2) = 0$

$x = 0,\ x = 2$ (critical values)

Also, $x \neq 1$ but this is not a critical value.

interval	$x < 0$	$0 < x < 1$	$1 < x < 2$	$x > 2$
sign of $f'(x)$	$+$	$-$	$-$	$+$
location	left of 0	right of 0 left of -1	right of 1 left of 2	right of 2

local maximum $(0, 0)$

local minimum $(2, 4)$

15. $f'(x) = 2x^2 - 2x - 12$ $f''(x) = 4x - 2$

Set $f'(x) = 0$: $f''(3) = 4(3) - 2 = 10$

$2x^2 - 2x - 12 = 0$ local minimum $(3, -25)$

$2(x^2 - x - 6) = 0$

$2(x - 3)(x + 2) = 0$ $f''(-2) = 4(-2) - 2 = -10$

$x = 3$ $x = -2$ (critical values)   local maximum $\left(-2, \frac{50}{3}\right)$

17. $f'(x) = 6x^2 + 27x - 15$ $f''(x) = 12x + 27$

Set $f'(x) = 0$: $f''\left(\frac{1}{2}\right) = 12\left(\frac{1}{2}\right) + 27 = 33$

$6x^2 + 27x - 15 = 0$ local minimum $\left(\frac{1}{2}, -1.875\right)$

$3(2x^2 + 9x - 5) = 0$

$3(2x - 1)(x + 5) = 0$ $f''(-5) = 12(-5) + 27 = -33$

$x = \frac{1}{2}$ $x = -5$ (critical values)   local maximum $(-5, 164.5)$

19. $f'(x) = 15x^4 - 60x^2$

Set $f'(x) = 0$: $15x^4 - 60x^2 = 0$, $15x^2(x^2 - 4) = 0$, $15x^2(x + 2)(x - 2) = 0$

                $x = 0$, $x = -2$, $x = 2$ (critical values)

$f''(x) = 60x^3 - 120x^2$

$f'(0) = 0$ This means we must use the first derivative test:

interval	$x < -2$	$-2 < x < 0$	$0 < x < 2$	$x > 2$
sign of $f'(x)$	$+$	$-$	$-$	$+$
location	left of -2	right of -2 left of 0	right of 0 left of 2	right of 2

local maximum $(-2, 64)$

local minimum $(2, -64)$

21a. $R(x) = px = (20 - 0.2x)x = 20x - 0.2x^2$

21b.

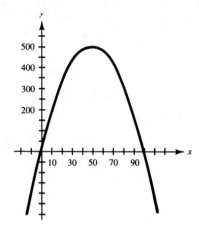

21c. $R'(x) = 20 - 0.4x$

Set $R'(x) = 0$: $20 - 0.4x = 0$

$$x = 50$$
(critical value)

Using the second derivative test:
$$R''(x) = -0.4$$

$R(x)$ is therefore a maximum at $x = 50$.

21d. $p = 20 - 0.2(50) = \$10$

21e. $R(50) = 20(50) - 0.2(50)^2 = \$500$

23. $P'(x) = -4x + 12$

Set $P'(x) = 0$: $-4x + 12 = 0$

$$x = 3$$
(critical value)

Using the second derivative test:
$$P''(x) = -4$$

$P(x)$ is therefore a maximum at $x = 3$

$$P(3) = -2(3)^2 + 12(3) - 7 = 11$$
$$(\$11,000)$$

25. $C'(x) = 12x^2 - 12x - 24$

To minimize $C'(x)$:

$$C''(x) = 24x - 12$$

Set $C''(x) = 0$: $24x - 12 = 0$

$$x = \frac{1}{2}$$
(critical value)

To use the second derivative test, find $C'''(x)$:
$$C'''(x) = 24$$

Therefore, the marginal cost $C'(x)$ is a minimum at $x = \frac{1}{2}$.

27. $R(x) = px = (48 - 4x^2)x = 48x - 4x^3$

$R'(x) = 48 - 12x^2$

Set $R'(x) = 0$: $48 - 12x^2 = 0$, $12(4 - x^2) = 0$, $12(2 + x)(2 - x) = 0$

Since x cannot be negative, $x = +2$ is a critical value.

Using the second derivative test: $R''(x) = -24x$
$$R''(2) = -24(2) = -48$$

The maximum revenue occurs at $x = 2$.

$$R(2) = 48(2) - 4(2)^3 = \$64, \ p = 48 - 4(2)^2 = \$32$$

29. $\dfrac{dN}{dx} = 120 - 2x$

Set $\dfrac{dN}{dx} = 0$: $120 - 2x = 0$, $x = 60$ (critical value)

Using the second derivative test: $\dfrac{d^2N}{dx^2} = -2$

The number of new customers is a maximum when $x = 60$. ($\$60,000$)

31. $\frac{dP}{dt} = 540t - 81t^2$

Set $\frac{dP}{dt} = 0$:

$$540t - 81t^2 = 0$$

$$9t(60 - 9t) = 0$$

$$t = 0 \qquad t = \frac{60}{9} = \frac{20}{3}$$

(critical values)

Using the second derivative test:

$$\frac{d^2P}{dt^2} = 540 - 162t$$

$$\left.\frac{d^2P}{dt^2}\right|_{t=0} = 540 - 162(0) = 540$$

$$\left.\frac{d^2P}{dt^2}\right|_{t=0} = 540 - 162\left(\frac{20}{3}\right) = -540$$

The maximum will occur at $t = \frac{20}{3}$
(6 years, 8 months)

33. $s'(t) = -32t + 64$

Set $s'(t) = 0$:

$$-32t + 64 = 0$$

$$t = 2 \text{ (critical value)}$$

Using the second derivative test:

$$s''(t) = -32$$

The height is a maximum at $t = 2$.

$$s(2) = -16(2)^2 + 64(2) = 64 \text{ ft.}$$

35. $v'(t) = \frac{(t^2+1)(10) - 10t(2t)}{(t^2+1)^2} = \frac{10 - 10t^2}{(t^2+1)^2} = \frac{10(1-t^2)}{(t^2+1)^2}$

Set $v'(t) = 0$: $\quad \frac{10(1-t^2)}{(t^2+1)^2} = 0, 10(1-t^2) = 0, 10(1+t)(1-t) = 0$

Since t cannot be negative, $t = 1$ is a critical value.

Using the first derivative test:

interval	$0 < t < 1$	$t > 1$
sign of $v'(t)$	$+$	$-$
location	left of 1	right of 1

The maximum velocity occurs when $t = 1$.

37. f_6 The function has a constant negative slope for $x < 0$ and a constant positive slope for $x > 0$.

39. f_5 The function is quadratic so its derivative must be a linear function.

41. f_3 The function decreases for $x < 0$ and increases for $x > 0$. There is a vertical asymptote, $x = 0$.

Section 4.4 (pages 263-265)

1.
$$\frac{dy}{dx} = -2x$$

Set $\frac{dy}{dx} = 0$: $-2x = 0$

$$x = 0$$

interval	$x < 0$	$x > 0$
sign of $\frac{dy}{dx}$	+	−
result: $f(x)$ is	increasing	decreasing

local maximum at $x = 0$

$f(0) = 4 - (0)^2 = 4$

absolute maximum value is 4

3.
$$f'(x) = -4(x+3)^3$$

Set $f'(x) = 0$: $-4(x+3)^3 = 0$

$$x = -3$$
(critical value)

interval	$x < -3$	$x > -3$
sign of $f'(x)$	+	−
result: $f(x)$ is	increasing	decreasing

local maximum at $x = -3$

$f(-3) = 1 - (-3+3)^4 = 1$

absolute maximum value is 1

5.
$$f'(x) = 3x^2 - 2x$$

Set $f'(x) = 0$: $3x^2 - 2x = 0$, $x(3x - 2) = 0$

$$x = 0, \ x = \tfrac{2}{3} \text{ (critical values)}$$

Using the second derivative test:

$f''(x) = 6x - 2$

$f''(0) = 6(0) - 2 = -2$ (local maximum at $x = 0$)

$f''\left(\tfrac{2}{3}\right) = 6\left(\tfrac{2}{3}\right) - 2 = 2$ (local minimum at $x = \tfrac{2}{3}$)

To find the absolute maximum, since $0 \le x \le 3$:

$f(0) = (0)^3 - (0)^2 = 0$, $f(3) = (3)^3 - (3)^2 = 18$

absolute maximum value is 18

7.
$$f'(x) = 6x^2 - 4x$$

Set $f'(x) = 0$: $6x^2 - 4x = 0$, $2x(3x - 2) = 0$

$$x = 0, \ x = \tfrac{2}{3} \text{ (critical values)}$$

interval	$x < 0$	$0 < x < \tfrac{2}{3}$	$\tfrac{2}{3} < x < 5$
sign of $f'(x)$	+	−	+
result: $f(x)$ is	increasing	decreasing	increasing

local maximum at $x = 0$

To find the absolute maximum: $f(0) = 2(0)^3 - 2(0)^2 + 5 = 5$

$f(5) = 2(5)^3 - 2(5)^2 + 5 = 205$

absolute maximum value is 205

9. $\dfrac{dy}{dx} = 2x - 2$

Set $\dfrac{dy}{dx} = 0$: $2x - 2 = 0$

$x = 1$ (critical value)

interval	$x < 1$	$x > 1$
sign of $\dfrac{dy}{dx}$	$-$	$+$
result: y is	decreasing	increasing

local minimum at $x = 1$

$f(1) = (1)^2 - 2(1) - 8 = -9$

absolute minimum value is -9

11. $y = -1(x^2 + 4)^{-1}$

$\dfrac{dy}{dx} = -1(-1)(x^2 + 4)^{-2}(2x)$

$\quad = \dfrac{2x}{(x^2 + 4)^2}$

Set $\dfrac{dy}{dx} = 0$: $\dfrac{2x}{(x^2 + 4)^2} = 0$

$2x = 0, \; x = 0$ (critical value)

interval	$x < 0$	$x > 0$
sign of $\dfrac{dy}{dx}$	$-$	$+$
result: y is	decreasing	increasing

local minimum at $x = 0$

$f(0) = \dfrac{-1}{(0)^2 + 4} = \dfrac{-1}{4}$

absolute minimum value is $\dfrac{-1}{4}$

13. $\dfrac{dy}{dx} = \dfrac{(x^2 + 1)(2x) - x^2(2x)}{(x^2 + 1)^2} = \dfrac{2x}{(x^2 + 1)^2}$

Set $\dfrac{dy}{dx} = 0$: $\dfrac{2x}{(x^2 + 1)^2} = 0, \; 2x = 0, \; x = 0$ (critical value)

interval	$x < 0$	$x > 0$
sign of $\dfrac{dy}{dx}$	$-$	$+$
result: y is	decreasing	increasing

local minimum at $x = 0$, $f(0) = \dfrac{(0)^2}{(0)^2 + 1} = 0$

absolute minimum value is 0

15. $\dfrac{dy}{dx} = 4x^3 - 10x$

Set $\dfrac{dy}{dx} = 0$: $4x^3 - 10x = 0, \; 2x(2x^2 - 5) = 0$

$x = 0, \; x = \pm\sqrt{2.5}$ (critical values)

interval	$x < -\sqrt{2.5}$	$-\sqrt{2.5} < x < 0$	$0 < x < \sqrt{2.5}$	$x > \sqrt{2.5}$
sign of $\dfrac{dy}{dx}$	$-$	$+$	$-$	$+$
result: y is	decreasing	increasing	decreasing	increasing

local minimum at $-\sqrt{2.5}$, $f(-\sqrt{2.5}) = \dfrac{-9}{4}$

local minimum at $\sqrt{2.5}$, $f(\sqrt{2.5}) = \dfrac{-9}{4}$

absolute minimum value is $\dfrac{-9}{4}$

17. $f'(x) = 6x - 3x^2$

   Set $f'(x) = 0$: $6x - 3x^2 = 0$, $3x(2-x) = 0$

   $\qquad\qquad\qquad x = 0 \qquad x = 2$ (critical values)

   Using the second derivative test:

   $\qquad f''(x) = 6 - 6x$
   $\qquad f''(0) = 6 - 6(0) = 6$ (local minimum at $x = 0$)
   $\qquad f''(2) = 6 - 6(2) = -6$ (local maximum at $x = 2$)

   To find the absolute maximum and minimum:

   $\qquad f(-0.5) = 3(-0.5)^2 - (-0.5)^3 + 1 = 1.875$
   $\qquad f(0) = 3(0)^2 - (0)^3 + 1 = 1$
   $\qquad f(2) = 3(2)^2 - (2)^3 + 1 = 5$ absolute maximum
   $\qquad f(5) = 3(5)^2 - (5)^3 + 1 = -49$ absolute minimum

19. $f'(x) = x^2 - x - 6$

   Set $f'(x) = 0$: $x^2 - x - 6 = 0$, $(x-3)(x+2) = 0$

   $\qquad\qquad\qquad x = 3 \qquad x = -2$ (critical values)

   Using the second derivative test:

   $\qquad f''(x) = 2x - 1$
   $\qquad f''(3) = 2(3) - 1 = 5$ (local minimum at $x = 3$)
   $\qquad f''(-2) = 2(-2) - 1 = -5$ (local maximum at $x = -2$)

   To find the absolute maximum and minimum:
   $$f(-3) = \frac{(-3)^3}{3} - \frac{(-3)^2}{2} - 6(-3) + 4 = 8\tfrac{1}{2}$$
   $$f(-2) = \frac{(-2)^3}{3} - \frac{(-2)^2}{2} - 6(-2) + 4 = 11\tfrac{1}{3} \text{ absolute maximum}$$
   $$f(3) = \frac{(3)^3}{3} - \frac{(3)^2}{2} - 6(3) + 4 = -9\tfrac{1}{2} \text{ absolute minimum}$$

21. $f'(x) = 24x - 12x^3$

   Set $f'(x) = 0$: $24x - 12x^3 = 0$, $12x(2 - x^2) = 0$

   $\qquad\qquad\qquad x = 0 \qquad x = \pm\sqrt{2}$ (critical values)

   Since $-1 \le x \le 1$, we do not need to check $x = \pm\sqrt{2}$.

   $\qquad f(-1) = 12(-1)^2 - 3(-1)^4 = 9$ absolute maximum
   $\qquad f(0) = 12(0)^2 - 3(0)^4 = 0$ absolute minimum
   $\qquad f(1) = 12(1)^2 - 3(1)^4 = 9$ absolute maximum

23.   From problem 21, the critical values are $x = 0, x = \pm\sqrt{2}$. To find the
absolute maximum and absolute minimum:

$$f(-2) = 12(-2)^2 - 3(-2)^4 = 0$$
$$f(-\sqrt{2}) = 12(-\sqrt{2})^2 - 3(-\sqrt{2})^4 = 12 \quad \text{absolute maximum}$$
$$f(0) = 0$$
$$f(\sqrt{2}) = 12(\sqrt{2})^2 - 3(\sqrt{2})^4 = 12 \quad \text{absolute maximum}$$
$$f(3) = 12(3)^2 - 3(3)^4 = -135 \quad \text{absolute minimum}$$

25. $g'(w) = 3w^2 - 4w + 1$

Set $g'(w) = 0$: $3w^2 - 4w + 1 = 0, (3w - 1)(w - 1) = 0$
$$w = \tfrac{1}{3} \qquad w = 1 \text{ (critical values)}$$

Using the second derivative test:

$$g''(w) = 6w - 4$$
$$g''\left(\tfrac{1}{3}\right) = 6\left(\tfrac{1}{3}\right) - 4 = -2 \text{ (local maximum at } w = \tfrac{1}{3})$$
$$g''(1) = 6(1) - 4 = 2 \text{ (local minimum at } w = 1)$$

To find the absolute maximum and minimum:

$$g(-2) = (-2)^3 - 2(-2)^2 + (-2) + 2 = -16 \quad \text{absolute minimum}$$
$$g\left(\tfrac{1}{3}\right) = \left(\tfrac{1}{3}\right)^3 - 2\left(\tfrac{1}{3}\right)^2 + \tfrac{1}{3} + 2 = 2\tfrac{4}{27}$$
$$g(1) = (1)^3 - 2(1)^2 + 1 + 2 = 2$$
$$g(3) = (3)^3 - 2(3)^2 + 3 + 2 = 14 \quad \text{absolute maximum}$$

27.   From problem 25, the critical values are $w = \tfrac{1}{3}, w = 1$.
$$g(-3) = (-3)^3 - 2(-3)^2 + (-3) + 2 = -46 \quad \text{absolute minimum}$$
$$g\left(\tfrac{1}{3}\right) = 2\tfrac{4}{27}$$
$$g(1) = 2$$
$$g(4) = (4)^3 - 2(4)^2 + 4 + 2 = 38 \quad \text{absolute maximum}$$

29.   There is a vertical asymptote at $x = 2$.

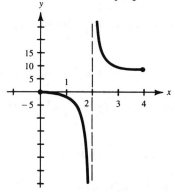

31a. $P(x) = 200x^{1/2} - x - 1000$

$P'(x) = 200\left(\frac{1}{2}\right)x^{-1/2} - 1 = \frac{100}{\sqrt{x}} - 1 = \frac{100 - \sqrt{x}}{\sqrt{x}}$

Set $P'(x) = 0$: $100 - \sqrt{x} = 0$, $x = \$10,000$ (critical value)

interval	$0 < x < 10,000$	$x > 10,000$
sign of $P'(x)$	+	−
result: $P(x)$ is	increasing	decreasing

absolute maximum: $P(10,000) = 200\sqrt{10,000} - 10,000 - 1000 = \$9000$

31b. $P(8000) = 200\sqrt{8000} - 8000 - 1000$

$\approx \$8888.54$

33. $\overline{C}(x) = \frac{2x^3 - 4x^2 + 4x}{x} = 2x^2 - 4x + 4$

$\overline{C}'(x) = 4x - 4$

Set $\overline{C}'(x) = 0$: $4x - 4 = 0$, $x = 1$ (critical value)

Since this is not in the given interval we only need to check the endpoints:

$\overline{C}(3) = 2(3)^2 - 4(3) + 4 = 10$

$\overline{C}(10) = 2(10)^2 - 4(10) + 4 = 164$

The absolute minimum occurs when $x = 3$.

Section 4.4A   (pages 269)

1.   From problem 17, section 4.4, there is a local minimum at $x = 0$ and a local maximum at $x = 2$.
From the graphing calculator the absolute minimum is $f(5) = -49$ and the absolute maximum is $f(-2) = 21$.

3. From problem 17, section 4.4, there is a local minimum at $x = 0$ and a local maximum at $x = 2$.
From the graphing calculator the absolute minimum is $f(0) = 1$ and the absolute maximum is $f(2) = 5$.

5. From problem 19, section 4.4 there is a local minimum at $x = 3$ and a local maximum at $x = -2$.
From the graphing calculator the absolute minimum is $f(3) = -9\frac{1}{2}$ and the absolute maximum is $f(-2) = 11\frac{1}{3}$.

7. From problems 21 and 23, section 4.4, there is a local minimum at $x = 0$ and local maxima at $x = \pm\sqrt{2}$.
From a graphing calculator the absolute minimum is $f(-2) = f(0) = f(2) = 0$ and the absolute maximum is $f(-\sqrt{2}) = f(\sqrt{2}) = 12$.

9. $S'(t) = 3(t-1)^2$
Set $S'(t) = 0$: $3(t-1)^2 = 0$, $t = 1$ (critical value)

interval	$t < 1$	$t > 1$
sign of $S'(t)$	+	+
result: $S(t)$ is	increasing	increasing

so there is no local maximum or minimum at $t = 1$. From a graphing calculator the absolute minimum is $f(-2) = -31$ and the absolute maximum is $f(2) = -3$.

11. From problems 25 and 27, section 4.4, there is a local minimum at $w = 1$ and a local maximum at $w = \frac{1}{3}$.
From a graphing calculator the absolute minimum is $f(-1) = -2$ and the absolute maximum is $f(1.5) = 2.375$.

13. From problems 25 and 27, section 4.4 there is a local minimum at $w = 1$ and a local maximum at $w = \frac{1}{3}$.
From a graphing calculator the absolute minimum is $f(-.5) = 0.875$ and the absolute maximum is $f(2.5) = 7.625$.

15. There is a vertical asymptote, $x = 2$.

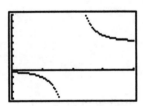

Section 4.5 (pages 278-281)

1. *y-intercept:* $\quad y = 0^2 + 1 \qquad$ *x-intercept:* none $\qquad$ *symmetry:* y-axis
   $y = 1$
   $(0,1)$

   critical values: $\quad \dfrac{dy}{dx} = 2x \qquad$ *concavity:* $\quad \dfrac{d^2y}{dx^2} = 2$
   $ 2x = 0 \qquad $ concave up for all x
   $ x = 0$

interval	$x < 0$	$x > 0$
sign of $\dfrac{dy}{dx}$	$-$	$+$
result: y is	decreasing	increasing

   local minimum at $(0,1)$

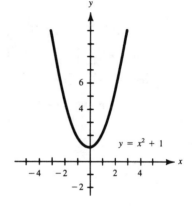

$y = x^2 + 1$

3. *y-intercept:* $\quad y = -3 \qquad$ *x-intercepts:* $\quad 2x^2 - 5x - 3 = 0 \qquad$ *symmetry:* none
   $ (0,-3) \qquad (2x+1)(x-3)$
   $ x = \dfrac{-1}{2} \qquad x = 3$
   $ \left(\dfrac{-1}{2}, 0\right) \qquad (3,0)$

   critical values: $\quad \dfrac{dy}{dx} = 4x - 5 \qquad$ *concavity:* $\quad \dfrac{d^2y}{dx^2} = 4$
   $ 4x - 5 = 0 \qquad $ concave up for all x
   $ x = \dfrac{5}{4}$

interval	$x < \dfrac{5}{4}$	$x > \dfrac{5}{4}$
sign of $\dfrac{dy}{dx}$	$-$	$+$
result: y is	decreasing	increasing

   local minimum at $\left(\dfrac{5}{4}, -6\dfrac{1}{8}\right)$

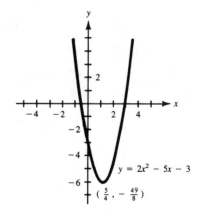

$y = 2x^2 - 5x - 3$

$\left(\dfrac{5}{4}, -\dfrac{49}{8}\right)$

5. *y-intercept:* $y = 6$ *x-intercepts:* $6 - x - x^2 = 0$ *symmetry:* none

         $(0, 6)$ $(3 + x)(2 - x) = 0$

                                   $x = -3$ $x = 2$

                                   $(-3, 0)$ $(2, 0)$

critical values: $\dfrac{dy}{dx} = -1 - 2x$ *concavity:* $\dfrac{d^2y}{dx^2} = -2$

                $-1 - 2x = 0$ concave down for all x

                     $x = \dfrac{-1}{2}$

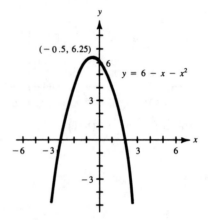

interval	$x < \frac{-1}{2}$	$x > \frac{-1}{2}$
sign of $\frac{dy}{dx}$	$+$	$-$
result: y is	increasing	decreasing

local maximum at $\left(\dfrac{-1}{2}, 6\dfrac{1}{4}\right)$

$(-0.5, 6.25)$

$y = 6 - x - x^2$

---

7. *y-intercept:* $y = (0 + 2)^3 = 8$ *x-intercepts:* $0 = (x + 2)^3$ *symmetry:*

            $(0, 8)$ $0 = x + 2$ none

                                     $-2 = x$

                                     $(-2, 0)$

critical values: $\dfrac{dy}{dx} = 3(x + 2)^2$ *concavity:* $\dfrac{d^2y}{dx^2} = 6(x + 2)$

             $3(x + 2)^2 = 0$ $6(x + 2) = 0$

                  $x = -2$ $x = -2$

interval	$x < -2$	$x > -2$
sign of $\frac{dy}{dx}$	$+$	$+$
result: y is	increasing	increasing

interval	$x < -2$	$x > -2$
sign of $\frac{d^2y}{dx^2}$	$-$	$+$
result: y is	concave down	concave up

point of inflection $(-2, 0)$

7. (continued)

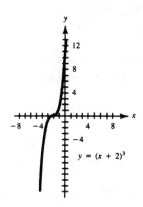

$y = (x + 2)^3$

9.   *y-intercept:* $f(0) = 15$ *symmetry:* none
        $(0, 15)$

critical values: $f'(x) = -24 + 30x - 6x^2$
                  $-24 + 30x - 6x^2 = 0, \; -6(x^2 - 5x + 4) = 0$
                  $-6(x - 1)(x - 4) = 0$
                  $x = 1 \quad\quad x = 4$

interval	$x < 1$	$1 < x < 4$	$x > 4$
sign of $f'(x)$	$-$	$+$	$-$
result: $f(x)$ is	decreasing	increasing	decreasing

local minimum at $(1, 4)$ *local maximum* at $(4, 31)$

concavity: $f''(x) = 30 - 12x$

           $30 - 12x = 0$

           $x = \dfrac{5}{2}$

interval	$x < \frac{5}{2}$	$x > \frac{5}{2}$
sign of $f''(x)$	$+$	$-$
result: $f(x)$ is	concave up	concave down

point of inflection $\left(\dfrac{5}{2}, 17\dfrac{1}{2}\right)$

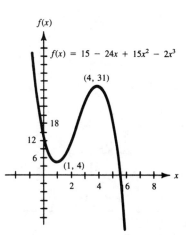

$f(x) = 15 - 24x + 15x^2 - 2x^3$

(4, 31)

18

12

6

(1, 4)

11.   *y-intercept:* $f(0) = 0$ *x-intercepts:* $x^3 - 3x^2 - 9x = 0$ *symmetry:* none
          $(0,0)$ $x(x^2 - 3x - 9) = 0$

$$x = 0 \qquad x \approx 4.85 \qquad x \approx -1.85$$

(from the quadratic formula)

critical values: $f'(x) = 3x^2 - 6x - 9$
$$3x^2 - 6x - 9 = 0, \; 3(x^2 - 2x - 3) = 0$$
$$3(x-3)(x+1) = 0$$
$$x = 3 \qquad x = -1$$

interval	$x < -1$	$-1 < x < 3$	$x > 3$
sign of $f'(x)$	+	−	+
result: $f(x)$ is	increasing	decreasing	increasing

local maximum at $(-1, 5)$ *local minimum at* $(3, -27)$

concavity: $f''(x) = 6x - 6$

interval	$x < 1$	$x > 1$
sign of $f''(x)$	−	+
result: $f(x)$ is	concave down	concave up

$$6x - 6 = 0$$

$$x = 1$$

point of inflection $(1, -11)$

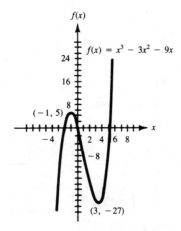

$f(x)$

$f(x) = x^3 - 3x^2 - 9x$

$(-1, 5)$

$(3, -27)$

13.   *y-intercept:* $f(0) = -1$ *symmetry:* none
          $(0, -1)$

critical values: $f'(x) = x^2 + 2x - 8$
$$x^2 + 2x - 8 = 0, \; (x+4)(x-2) = 0$$
$$x = -4 \qquad x = 2$$

interval	$x < -4$	$-4 < x < 2$	$x > 2$
sign of $f'(x)$	+	−	+
result: $f(x)$ is	increasing	decreasing	increasing

local maximum at $\left(-4, \frac{77}{3}\right)$ *local minimum at* $\left(2, \frac{-31}{3}\right)$

concavity: $f''(x) = 2x + 2$

interval	$x < -1$	$x > -1$
sign of $f''(x)$	−	+
result: $f(x)$ is	concave down	concave up

$$2x + 2 = 0$$

$$x = -1$$

point of inflection $\left(-1, \frac{23}{3}\right)$

13.   (continued)

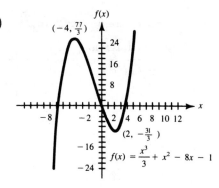

$(-4, \frac{77}{3})$

$(2, -\frac{31}{3})$

$f(x) = \frac{x^3}{3} + x^2 - 8x - 1$

15.   *y-intercept:* $y = 0$ *x-intercepts:* $4x - x^4 = 0$ *symmetry:* none
            $(0,0)$ $x(4 - x^3) = 0$
                                                       $x = 0 \quad x = \sqrt[3]{4}$

critical values: $\dfrac{dy}{dx} = 4 - 4x^3$ *concavity:* $\dfrac{d^2y}{dx^2} = -12x^2$

$ 4 - 4x^3 = 0,\ 4(1 - x^3) = 0$ $-12x^2 = 0$

$ x = 1$ $x = 0$

interval	$x < 1$	$x > 1$
sign of $\dfrac{dy}{dx}$	$+$	$-$
result: y is	increasing	decreasing

interval	$x < 0$	$x > 0$
sign of $\dfrac{d^2y}{dx^2}$	$-$	$-$
result: y is	concave down	concave down

local maximum at $(1, 3)$

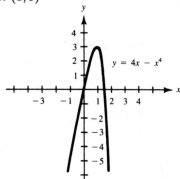

$y = 4x - x^4$

17.   *y-intercept:* $y = 0$ *x-intercepts:* $12x - x^3 = 0$ *symmetry:* origin
            $(0,0)$ $x(12x^2 - x^2) = 0$
                                                       $x = 0 \quad x = \pm 2\sqrt{3}$
                                                       $(0,0) \quad (2\sqrt{3}, 0) \quad (-2\sqrt{3}, 0)$

critical values: $f'(x) = 12 - 3x^2$
$ 12 - 3x^2 = 0,\ 3(4 - x^2) = 0$
$ x = \pm 2$

interval	$x < -2$	$-2 < x < 2$	$x > 2$
sign of $f'(x)$	$-$	$+$	$-$
result: $f(x)$ is	decreasing	increasing	decreasing

local minimum at $(-2,-16)$ local maximum at $(2,16)$

concavity: $f''(x) = -6x$

$-6x = 0$

$x = 0$

interval	$x < 0$	$x > 0$
sign of $f''(x)$	$+$	$-$
result: $f(x)$ is	concave up	concave down

point of inflection $(0,0)$

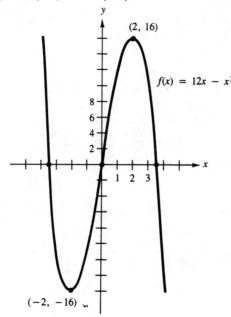

$(2,16)$

$f(x) = 12x - x^3$

$(-2,-16)$

19.   y-intercept: $y = 0$ x-intercepts: $x^2 - 2x^4 = 0$ symmetry: y-axis
$(0,0)$ $x^2(1 - 2x^2) = 0$

$$x = 0 \qquad x = \pm\frac{\sqrt{2}}{2}$$

$$(0,0) \qquad \left(\frac{\sqrt{2}}{2}, 0\right) \qquad \left(-\frac{\sqrt{2}}{2}, 0\right)$$

critical values: $\dfrac{dy}{dx} = 2x - 8x^3$

$2x - 8x^3 = 0, \ 2x(1 - 4x^2) = 0$

$2x(1 + 2x)(1 - 2x) = 0$

$$x = 0 \qquad x = \frac{-1}{2} \qquad x = \frac{1}{2}$$

interval	$x < \frac{-1}{2}$	$\frac{-1}{2} < x < 0$	$0 < x < \frac{1}{2}$	$x > \frac{1}{2}$
sign of $\dfrac{d^2y}{dx^2}$	$+$	$-$	$+$	$-$
result: y is	increasing	decreasing	increasing	decreasing

local maximum at $\left(\frac{-1}{2}, \frac{1}{8}\right)$ and $\left(\frac{+1}{2}, \frac{1}{8}\right)$ local minimum at $(0,0)$

$concavity:$ $\dfrac{d^2y}{dx^2} = 2 - 24x^2,\ 2 - 24x^2 = 0,\ 2(1 - 12x^2) = 0,\ x = \pm\dfrac{\sqrt{3}}{6}$

interval	$x < \dfrac{-\sqrt{3}}{6}$	$\dfrac{-\sqrt{3}}{6} < x < \dfrac{\sqrt{3}}{6}$	$x > \dfrac{\sqrt{3}}{6}$
sign of $\dfrac{d^2y}{dx^2}$	$-$	$+$	$-$
result: y is	concave down	concave up	concave down

$points\ of\ inflection:$ $\left(-\dfrac{\sqrt{3}}{6}, \dfrac{5}{72}\right)$ $\left(-\dfrac{\sqrt{3}}{6}, \dfrac{5}{72}\right)$

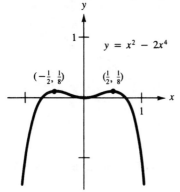

21. no

23. none

25. $f'(x)$ is positive when the graph is increasing: $-2 < x < 0,\ x > 3.5$
$f'(x)$ is negative when the graph is decreasing: $x < -2,\ 0 < x < 3.5$

27. absolute minimum: -55

29. symmetric with respect to the y-axis
y-intercept $\quad (0,1)$
vertical asymptotes: $\quad x = -1,\ x = 1$
horizontal asymptote: $\quad y = 0$

31. none

33. y-intercept: none $\qquad$ x-intercept: $\dfrac{1}{x} + 4 = 0$ $\qquad$ symmetry: none

$critical\ values:$ $\quad f(x) = x^{-1} + 4 \qquad \dfrac{1}{x} = -4$

$\qquad\qquad\qquad\quad f'(x) = -1x^{-2} \qquad x = \dfrac{-1}{4}$

$\qquad\qquad\qquad\qquad\quad = \dfrac{1}{x^2} \qquad\qquad \left(\dfrac{-1}{4}, 0\right)$

There are no critical values but $f(x)$ is undefined at $x = 0$.

interval	$x < 0$	$x > 0$
sign of $f'(x)$	$-$	$-$
result: $f(x)$ is	decreasing	decreasing

$concavity:$ $\quad f''(x) = (-1)(-2)x^{-3}$

$\qquad\qquad\qquad\quad = \dfrac{2}{x^3}$

interval	$x < 0$	$x > 0$
sign of $f'(x)$	$-$	$+$
result: $f(x)$ is	concave down	concave up

vertical asymptote: $x = 0$ since $\lim\limits_{x \to 0^-} f(x) = -\infty$ and $\lim\limits_{x \to 0^+} f(x) = +\infty$

horizontal asymptote: $y = 4$ since $\lim\limits_{x \to \infty} (\frac{1}{x} + 4) = 0 + 4 = 4$

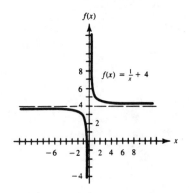

$f(x) = \frac{1}{x} + 4$

35. *y-intercept:* $f(0) = \dfrac{1}{0-4} = \dfrac{-1}{4}$ *x-intercept:* none   *symmetry:* y-axis

$(0, \dfrac{-1}{4})$

critical values: $f(x) = (x^2 - 4)^{-1}$

$$f'(x) = -1(x^2-4)^{-2}(2x) = \frac{-2x}{(x^2-4)^2}$$

$$\frac{-2x}{(x^2-4)^2} = 0$$

$$-2x = 0, \ x = 0$$

$f(x)$ is not defined at $x = \pm 2$ (but these are not ciritcal values)

interval	$x < -2$	$-2 < x < 0$	$0 < x < 2$	$x > 2$
sign of $f'(x)$	$+$	$+$	$-$	$-$
result: $f(x)$ is	increasing	increasing	decreasing	decreasing

local maximum at $(0, \dfrac{-1}{4})$

concavity: $\dfrac{(x^2-4)^2(-2) - (-2x)(2)(x^2-4)(2x)}{[(x^2-4)^2]^2} = \dfrac{-2(x^2-4)[(x^2-4) - 2x(2x)]}{(x^2-4)^4}$

$$= \frac{2(3x^2+4)}{(x^2-4)^3}$$

interval	$x < -2$	$-2 < x < 2$	$x > 2$
sign of $f''(x)$	$+$	$-$	$+$
result: $f(x)$ is	concave up	concave down	concave up

vertical asymptotes: $x = 2, x = -2$ since $\lim\limits_{x \to -2^-} f(x) = +\infty$ and $\lim\limits_{x \to -2^+} f(x) = -\infty$

$$\lim\limits_{x \to 2^-} f(x) = -\infty \text{ and } \lim\limits_{x \to 2^+} f(x) = +\infty$$

horizontal asymptote: $y = 0$ since $\lim\limits_{x \to \infty} \dfrac{1}{x^2 - 4} = \lim\limits_{x \to \infty} \dfrac{x^2}{1 - \dfrac{4}{x^2}}$

$$= \dfrac{0}{1 - 0} = 0$$

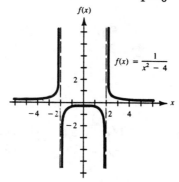

37. *y-intercept:* $y = \dfrac{1}{(0+2)^2} = \dfrac{1}{4}$ *x-intercept:* none   *symmetry:* none

$\left(0, \dfrac{1}{4}\right)$

critical values: $y = (x+2)^{-2}$

$$\dfrac{dy}{dx} = -2(x+2)^{-3} = \dfrac{-2}{(x+2)^3}$$

no critical values but $f(x)$ is undefined for $x = -2$

interval	$x < -2$	$x > -2$
sign of $\dfrac{dy}{dx}$	$+$	$-$
result: y is	increasing	decreasing

concavity: $\dfrac{d^2y}{dx^2} = -2(-3)(x+2)^{-4} = \dfrac{6}{(x+2)^4}$

interval	$x < -2$	$x > -2$
sign of $\dfrac{d^2y}{dx^2}$	$+$	$+$
result: y is	concave up	concave up

vertical asymptote: $x = -2$ since $\lim\limits_{x \to -2^-} f(x) = +\infty$ and $\lim\limits_{x \to -2^+} f(x) = +\infty$

horizontal asymptote: $y = 0$ since $\lim\limits_{x \to \infty} \dfrac{1}{(x+2)^2}$

$$= \lim\limits_{x \to \infty} \dfrac{1}{x^2 + 4x + 4} = \lim\limits_{x \to \infty} \dfrac{\dfrac{1}{x^2}}{1 + \dfrac{4}{x} + \dfrac{4}{x^2}} = \dfrac{0}{1 + 0 + 0} = 0$$

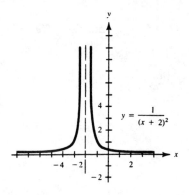

$y = \dfrac{1}{(x+2)^2}$

39. *y-intercept:* $y = 1 + \sqrt[3]{0-1} = 0$ *x-intercepts:* $1 + \sqrt[3]{x-1} = 0$ *symmetry:*
    $(0,0)$ $\sqrt[3]{x-1} = -1$ none

    $x - 1 = -1$

    $x = 0$

    $(0,0)$

critical values:
$$y = 1 + (x-1)^{1/3}$$
$$\frac{dy}{dx} = \tfrac{1}{3}(x-1)^{-2/3} = \frac{1}{3\sqrt[3]{(x-1)^2}}$$
$\dfrac{dy}{dx}$ is undefined for $x = 1$

interval	$0 < x < 1$	$x > 1$
sign of $\dfrac{dy}{dx}$	$+$	$+$
result: y is	increasing	increasing

concavity:
$$\frac{d^2y}{dx^2} = \tfrac{1}{3}\left(\tfrac{-2}{3}\right)(x-1)^{-5/3} = \frac{-2}{9\sqrt[3]{(x-1)^5}}$$

$\dfrac{d^2y}{dx^2}$ is undefined for $x = 1$

interval	$0 < x < 1$	$x > 1$
sign of $\dfrac{d^2y}{dx^2}$	$+$	$-$
result: y is	concave up	concave down

point of inflection $(1,1)$

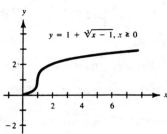

$y = 1 + \sqrt[3]{x-1},\, x \ge 0$

41.   *y-intercept:* $y = 0$ *x-intercepts:* $x^{4/3} + x^{1/3} = 0$ *symmetry:* none

$$x^{1/3}(x + 1) = 0$$

$$x = 0 \qquad x = -1 \text{ (not in domain)}$$

critical values: $f'(x) = \frac{4}{3}x^{1/3} + \frac{1}{3}x^{-2/3}$

$$= \frac{4\sqrt[3]{x}}{3} + \frac{1}{3\sqrt[3]{x^2}} = \frac{4x + 1}{3\sqrt[3]{x^2}}$$

$$\frac{4x + 1}{3\sqrt[3]{x^2}} = 0, \ 4x + 1 = 0, \ x = \frac{-1}{4} \text{ (not in domain)}$$

$f'(x)$ is undefined for $x = 0$.

interval	$x > 0$
sign of $f'(x)$	$+$
result: $f(x)$ is	increasing

concavity: $f''(x) = \frac{4}{3}\left(\frac{1}{3}\right)x^{-2/3} + \frac{1}{3}\left(\frac{-2}{3}\right)x^{-5/3} = \frac{4}{9\sqrt[3]{x^2}} - \frac{2}{9\sqrt[3]{x^5}} = \frac{4x - 2}{9\sqrt[3]{x^5}}$

$$\frac{4x - 2}{9\sqrt[3]{x^5}} = 0, \ 4x - 2 = 0, \ x = \frac{1}{2}$$

interval	$0 < x < \frac{1}{2}$	$x > \frac{1}{2}$
sign of $f''(x)$	$-$	$+$
result: $f(x)$ is	concave down	concave up

point of inflection $\left(\frac{1}{2}, 1.19\right)$

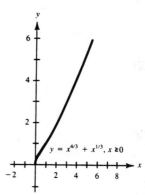

$y = x^{4/3} + x^{1/3}, \ x \geq 0$

43.   *critical values:* $P'(x) = -6x^2 + 150$

$$-6x^2 + 150 = 0$$
$$-6x^2 = -150$$
$$x^2 = 25$$
$$x = 5 \text{ since } 0 \leq x \leq 10$$

interval	$0 < x < 5$	$5 < x < 10$
sign of $P'(x)$	$+$	$-$
result: $P(x)$ is	increasing	decreasing

local maximum at $(5, 560)$

concavity: $P''(x) = -12x$
$$-12x = 0$$
$$x = 0$$

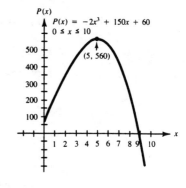

$P(x) = -2x^3 + 150x + 60$
$0 \leq x \leq 10$
$(5, 560)$

Since $0 \leq x \leq 10$, $P(x)$ is concave down for all x in the interval.

45a. $P(x) = R(x) - C(x)$
$\qquad = (-8x^3 - 4x^2 + 52x) - (2x^2 + 16x)$
$\qquad = -8x^3 - 6x^2 + 36x$

critical values: $P'(x) = -24x^2 - 12x + 36$
$\qquad\qquad -24x^2 - 12x + 36 = 0$
$\qquad\qquad -12(2x^2 + x - 3) = 0$
$\qquad\qquad -12(2x + 3)(x - 1) = 0$
$\qquad\qquad x = \dfrac{-3}{2} \qquad x = 1 \text{ (critical values)}$

interval	$x < \frac{-3}{2}$	$\frac{-3}{2} < x < 1$	$x > 1$
sign of $P'(x)$	$-$	$+$	$-$
result: $P(x)$ is	decreasing	increasing	decreasing

local minimum at $(\frac{-3}{2}, -40.5)$ $\qquad$ *local maximum at* $(1, 22)$

concavity: $P''(x) = -48x - 12$

$\qquad\qquad -48x - 12 = 0$

$\qquad\qquad x = -\dfrac{1}{4}$

interval	$x < \frac{-1}{4}$	$x > \frac{-1}{4}$
sign of $P''(x)$	$+$	$-$
result: $P(x)$ is	concave up	concave down

point of inflection $(\frac{-1}{4}, -9.25)$

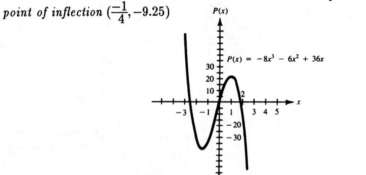

$P(x) = -8x^3 - 6x^2 + 36x$

45b. $C'(x) = 4x + 16$
$R'(x) = -24x^2 - 8x + 52$

critical value for marginal revenue: $R''(x) = -48x - 8$
$\qquad\qquad\qquad\qquad\qquad\qquad -48x - 8 = 0$

$\qquad\qquad\qquad\qquad\qquad\qquad x = \dfrac{-1}{6}$

local maximum at $\left(\dfrac{-1}{6}, 52\dfrac{2}{3}\right)$

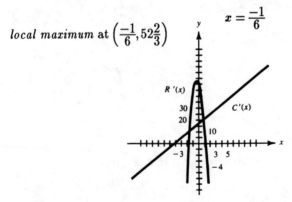

45c. $R'(x) = C'(x)$

$-24x^2 - 8x + 52 = 4x + 16$

$-24x^2 - 12x + 36 = 0$

$-12(2x^2 + x - 3) = 0$

$-12(2x + 3)(x - 1) = 0$

$x = \frac{-3}{2} \qquad x = 1$

Section 4.6 (pages 293-298)

1. $x + y = 160$

$y = 160 - x$

Maximize the product:

$P = xy$

$P = x(160 - x)$

$P = 160x - x^2$

$\frac{dP}{dx} = 160 - 2x$

Set $\frac{dP}{dx} = 0$:

$160 - 2x = 0$

$x = 80$

Using the second derivative test:

$\frac{d^2P}{dx^2} = -2$ (maximum)

$y = 160 - x = 160 - 80 = 80$

The numbers are 80 and 80.

3. $xy = 16$

$y = \frac{16}{x}$

Minimize the sum of their squares:

$S = x^2 + y^2$

$S = x^2 + \left(\frac{16}{x}\right)^2$

$= x^2 + 256x^{-2}$

$\frac{dS}{dx} = 2x - 512x^{-3}$

Set $\frac{dS}{dx} = 0$:

$2x - \frac{512}{x^3} = 0$

$2x^4 = 512$

$x^4 = 256, \ x = \pm 4$

Using the second derivative test:

$\frac{d^2S}{dx^2} = 2 + 1536x^{-4} = 2 + \frac{1536}{x^4}$

For $x = 4$ or $x = -4$

$\frac{d^2S}{dx^2} = 2 + 6 = 8$ (minimum)

The numbers are 4 and 4 or −4 and −4.

5. $xy = 48$

$y = \frac{48}{x}$

Minimize the sum of the first and 3 times the second:

$S = x + 3y$

$S = x + 3\left(\frac{48}{x}\right) = x + 144x^{-1}$

$\frac{dS}{dx} = 1 - 144x^{-2}$

Set $\frac{dS}{dx} = 0$: $\quad 1 - \frac{144}{x^2} = 0, \ x^2 = 144, \ x = 12$

Using the second derivative test: $\quad \frac{d^2S}{dx^2} = 288x^{-3}$

For $x = 12 \qquad \frac{d^2S}{dx^2} = \frac{288}{1728} = \frac{1}{6}$ (minimum) $\qquad$ The numbers are 12 and 4.

7. $A = lw$, $64 = xy$, $y = \frac{64}{x}$

Minimize the perimeter:

$P = 2x + 2y$

$P = 2x + 2\left(\frac{64}{x}\right)$

$P = 2x + 128x^{-1}$

$\frac{dP}{dx} = 2 - 128x^{-2}$

Set $\frac{dP}{dx} = 0$: $2 - \frac{128}{x^2} = 0$

$\qquad\qquad 2x^2 = 128$

$\qquad\qquad x^2 = 64$

$\qquad\qquad x = 8$

Using the second derivative test:

$\frac{d^2P}{dx^2} = 256x^{-3}$

For $x = 8$

$\frac{d^2P}{dx^2} = \frac{256}{512} = \frac{1}{2}$ (minimum)

The dimensions are 8 cm. and 8 cm.

9. $x^2 + y^2 = 12^2$, $y^2 = 144 - x^2$, $y = \sqrt{144 - x^2}$

Maximize the area:

$A = xy$

$A = x\sqrt{144 - x^2}$

$A = x(144 - x^2)^{1/2}$

$\frac{dA}{dx} = x\left(\frac{1}{2}\right)(144 - x^2)^{-1/2}(-2x) + (1)(144 - x^2)^{1/2}$

$\qquad = (144 - x^2)^{-1/2}[-x^2 + (144 - x^2)]$

$\qquad = \frac{144 - 2x^2}{\sqrt{144 - x^2}}$

Set $\frac{dA}{dx} = 0$: $\frac{144 - 2x^2}{\sqrt{144 - x^2}} = 0$

$\qquad\qquad 144 - 2x^2 = 0$

$\qquad\qquad -2x^2 = -144$

$\qquad\qquad x^2 = 72$

$\qquad\qquad x = \sqrt{72} = 6\sqrt{2}$

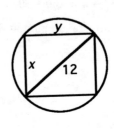

Using the first derivative test:

interval	$0 < x < 6\sqrt{2}$	$6\sqrt{2} < x < 12$
sign of $\frac{dA}{dx}$	$+$	$-$
location	right of 0 left of $6\sqrt{2}$	right of $6\sqrt{2}$ left of 12

A local maximum occurs at $x = 6\sqrt{2}$.

When $x = 6\sqrt{2}$, $y = \sqrt{144 - (6\sqrt{2})^2} = 6\sqrt{2}$

11.   length + width = 10 $x + y = 10$

$$y = 10 - x$$

Maximize the volume:

Volume = (length)(width)(height)

$V(x) = x(10 - x)(3)$

$V(x) = 30x - 3x^2$

$V'(x) = 30 - 6x$

Set $V'(x) = 0$: $30 - 6x = 0$

$$x = 5$$

Using the second derivative test:

$V''(x) = -6$ (maximum)

The length is 5 in.

The width is $10 - 5 = 5$ in.

The volume is $(5)(5)(3) = 75$ in.^{3}

---

13. $R(x) = px = (160 - 0.2x)x = 160x - 0.2x^2$

Maximize the revenue:

$R'(x) = 160 - 0.4x$

Set $R'(x) = 0$: $160 - 0.4x = 0$

$$x = 400$$

Using the second derivative test:

$R''(x) = -0.4$ (maximum)

The revenue is maximized when

400 rackets are sold.

---

15. $R(x) = 5x$

$P(x) = R(x) - C(x)$

$\quad = 5x - (500 - 118x - 36.5x^2 + 2x^3)$

$\quad = -500 + 123x + 36.5x^2 - 2x^3$

Maximize the profit:

$P'(x) = 123 + 73x - 6x^2$

Set $P'(x) = 0$: $123 + 73x - 6x^2 = 0$

$$(41 - 3x)(3 + 2x) = 0$$

$$x = \frac{41}{3} \qquad x = \frac{-3}{2}$$

Using the second derivative test:

$P''(x) = 73 - 12x$

For $x = \frac{41}{3}$, $P\left(\frac{41}{3}\right) = 73 - 12\left(\frac{41}{3}\right) = -91$ (maximum)

The maximum profit occurs for $x = \frac{41}{3}$ (approximately 13,667)

17. $\overline{C}(x) = \dfrac{0.005x^2 + 0.01x + 200}{x} = 0.005x + 0.01 + 200x^{-1}$

Minimize the average cost:

$\overline{C}'(x) = 0.005 - 200x^{-2}$

Set $\overline{C}'(x) = 0$: $\qquad 0.005 - \dfrac{200}{x^2} = 0$

$$0.05x^2 = 200$$

$$x^2 = 40{,}000$$

$$x = 200$$

Using the second derivative test:

$\overline{C}''(x) = 400x^{-3}$

$\overline{C}''(200) = \dfrac{1}{20000}$ (minimum)

The minimum occurs when 200 toasters are produced.

19. Minimize the cost:

$C'(x) = -4 + 0.016x$

Set $C'(x) = 0$: $\qquad -4 + 0.016x = 0$

$$x = 250$$

Using the second derivative test:

$C''(x) = 0.016$ (minimum)

The minimum occurs when 250 are produced.

21. Maximize the concentration:

$\dfrac{dC}{dt} = \dfrac{(t^2 + 3)(5) - 5t(2t)}{(t^2 + 3)^2} = \dfrac{15 - 5t^2}{(t^2 + 3)^2} = \dfrac{5(3 - t^2)}{(t^2 + 3)^2}$

Set $\dfrac{dC}{dt} = 0$: $\qquad \dfrac{5(3 - t^2)}{(t^2 + 3)^2} = 0$

$$5(3 - t^2) = 0$$

$$t = \sqrt{3} \text{ since } t \geq 0$$

Using the first derivative test:

interval	$0 < t < \sqrt{3}$	$x < \sqrt{3}$
sign of $\dfrac{dC}{dt}$	$+$	$-$
location	right of 0 left of $\sqrt{3}$	right of $\sqrt{3}$

The maximum occurs at $t = \sqrt{3}$ hours.

23. Maximize the profit:

$$\frac{dP}{dx} = 38 - 0.2x$$

Set $\frac{dP}{dx} = 0$: $\quad\quad 38 - 0.2x = 0$

$$x = 190$$

Using the second derivative test:

$$\frac{d^2P}{dx^2} = -0.2 \text{ (maximum)}$$

The profit is a maximum when 190 units are sold.

25. $V = \pi r^2 h,\ 64\pi = \pi r^2 h,\ h = \dfrac{64\pi}{\pi r^2} = \dfrac{64}{r^2}$

Minimize the surface area:

$$S = 2\pi r^2 + 2\pi r h$$

$$S = 2\pi r^2 + 2\pi r\left(\frac{64}{r^2}\right)$$

$$S = 2\pi r^2 + 128\pi r^{-1}$$

$$\frac{dS}{dr} = 4\pi r - 128\pi r^{-2}$$

Set $\frac{dS}{dr} = 0$: $\quad 4\pi r - \dfrac{128\pi}{r^2} = 0$

$$\frac{4\pi r^3 - 128\pi}{r^2} = 0$$

$$4\pi r^3 - 128\pi = 0$$

$$r^3 = 32$$

$$r = \sqrt[3]{32}$$

Using the second derivative test:

$$\frac{d^2S}{dr^2} = 4\pi + 256\pi r^{-3}$$

For $r = \sqrt[3]{32}$,

$$\frac{d^2S}{dr^2} = 4\pi + \frac{256\pi}{32} = 12\pi \text{ (minimum)}$$

When $r = \sqrt[3]{32}$, $h = \dfrac{64}{\left(\sqrt[3]{64}\right)^2}$

$$= \frac{64\sqrt[3]{32}}{32} = 2\sqrt[3]{32}$$

27. Volume = (length)(width)(height)

$$160 = x \cdot x \cdot y$$

$$160 = x^2 y$$

$$\frac{160}{x^2} = y$$

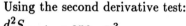

Minimize the cost:

Total cost = (cost of 4 sides)+(cost of top)+(cost of bottom)

$$C = 0.08(4)(xy) + 0.08x^2 + 0.12x^2$$

$$C = 0.32x\left(\frac{160}{x^2}\right) + 0.20x^2$$

$$C = 51.2x^{-1} + 0.20x^2$$

$$\frac{dC}{dx} = -51.2x^{-2} + 0.40x$$

Set $\dfrac{dC}{dx} = 0$: $\quad \dfrac{-51.2}{x^2} + 0.40x = 0$

$$0.40x^3 = 51.2$$

$$x^3 = 128$$

$$x = \sqrt[3]{128} = 4\sqrt[3]{2}$$

Using the second derivative test:

$\dfrac{d^2C}{dx^2} = 102.4x^{-3} + 0.40$

For $x = 4\sqrt[3]{2}$, $\dfrac{d^2C}{dx^2} = \dfrac{102.4}{128} + 0.40$ (minimum)

When $x = 4\sqrt[3]{2}$, $y = \dfrac{160}{(4\sqrt[3]{2})^2} = \dfrac{10}{4\sqrt[3]{2^2}} = 5\sqrt[3]{2}$

29. Let $y =$ the width of the auditorium
so $\frac{1}{2}y =$ the radius of the semicircular part
Let $x =$ the length of the building
The distance around the semicircular end is
half the circumference

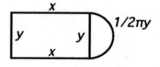

$$C = 2\pi r = 2\pi\left(\tfrac{1}{2}y\right) = \pi y$$

$$\tfrac{1}{2}C = \tfrac{1}{2}\pi y$$

The perimeter of the auditorium is

$P = 2x + y + \frac{1}{2}\pi y$

$500 = 2x + y + \frac{1}{2}\pi y$

$500 - 2x = y + \frac{1}{2}\pi y$

$500 - 2x = (1 + \frac{1}{2}\pi)y$

$y = \dfrac{500 - 2x}{1 + \frac{1}{2}\pi} = \dfrac{1000 - 4x}{2 + \pi}$

Maximize the area of the rectangular part:

$A = xy$

$A = x\left(\dfrac{1000 - 4x}{2 + \pi}\right)$

$A = \dfrac{1}{2 + \pi}(1000x - 4x^2)$

$\dfrac{dA}{dx} = \dfrac{1}{2 + \pi}(1000 - 8x)$

Set $\dfrac{dA}{dx} = 0$: $\quad \dfrac{1}{2 + \pi}(1000x - 8x) = 0$

$$1000 - 8x = 0$$

$$x = 125$$

Using the second derivative test:

$\dfrac{d^2A}{dx^2} = \dfrac{1}{2 + \pi}(-8)$ (maximum)   When $x = 125$, $y = \dfrac{1000 - 4(125)}{2 + \pi} = \dfrac{500}{2 + \pi}$

31. $d^2 + w^2 = 4^2$
$d^2 = 16 - w^2$

Maximize the strength of the beam:

$S = wd^2$
$S = w(16 - w^2)$
$S = 16w - w^3$

$\dfrac{dS}{dw} = 16 - 3w^2$

Set $\dfrac{dS}{dw} = 0$: $16 - 3w^2 = 0$

$$w^2 = \frac{16}{3}$$
$$w = \sqrt{\frac{16}{3}} = \frac{4}{\sqrt{3}} = \frac{4\sqrt{3}}{3}$$

Using the second derivative test:

$\dfrac{d^2S}{dw^2} = -6w$

When $w = \dfrac{4\sqrt{3}}{3}$, $\dfrac{d^2S}{dw^2} = -6\left(\dfrac{4\sqrt{3}}{3}\right) = -8\sqrt{3}$ (maximum)

$$d^2 = 16 - \left(\frac{4\sqrt{3}}{3}\right)^2 = 16 - \frac{16}{3} = \frac{32}{3}$$

$$d = \sqrt{\frac{32}{3}} = \frac{4\sqrt{2}}{\sqrt{3}} = \frac{4\sqrt{6}}{3}$$

33.   Let x = distance from A
$20 - x$ = distance from B
Total concentration = (pollution from A) + (pollution from B)

$$C = 2\left(\frac{1}{x^2}\right) + \frac{1}{(20-x)^2}$$

$$C = 2x^{-2} + (20-x)^{-2}$$

$$\frac{dC}{dx} = -4x^{-3} + (-2)(20-x)^{-3}(-1)$$

$$= \frac{-4}{x^3} + \frac{2}{(20-x)^3} = \frac{-4(20-x)^3 + 2x^3}{x^3(20-x)^3}$$

Set $\dfrac{dC}{dx} = 0$: $\dfrac{-4(20-x)^3 + 2x^3}{x^3(20-x)^3} = 0$

$$-4(20-x)^3 + 2x^3 = 0$$

$$2x^3 = 4(20-x)^3$$

$$x^3 = 2(20-x)^3$$

$$x = \sqrt[3]{2}(20-x)$$

$$x = 20\sqrt[3]{2} - \sqrt[3]{2}x$$

$$\sqrt[3]{2}x + x = 20\sqrt[3]{2}$$

$$(\sqrt[3]{2} + 1)x = 20\sqrt[3]{2} \qquad x = \frac{20\sqrt[3]{2}}{\sqrt[3]{2} + 1}$$

Using the first derivative test:

interval	$0 < x < \dfrac{20\sqrt[3]{2}}{\sqrt[3]{2}+1}$	$\dfrac{20\sqrt[3]{2}}{\sqrt[3]{2}+1} < x < 20$
sign of $\dfrac{dC}{dx}$	$-$	$+$
location	left of $\dfrac{20\sqrt[3]{2}}{\sqrt[3]{2}+1}$	right of $\dfrac{20\sqrt[3]{2}}{\sqrt[3]{2}+1}$

The concentration is a minimum at a distance of $\dfrac{20\sqrt[3]{2}}{\sqrt[3]{2}+1}$ miles from A.

35. $\dfrac{dS}{dt} = S_0(a + 2bt + 3ct^2)$

$= S_0[(5.3 \times 10^{-5}) + 2(-6.5 \times 10^{-6})t + 3(1.4 \times 10^{-8})t^2]$

$= S_0[(5.3 \times 10^{-5}) - (1.3 \times 10^{-5})t + (4.2 \times 10^{-8})t^2]$

Set $\dfrac{dS}{dt} = 0$: $S_0[(5.3 \times 10^{-5}) - (1.3 \times 10^{-5})t + (4.2 \times 10^{-8})t^2] = 0$

$(4.2 \times 10^{-8})t^2 - (1.3 \times 10^{-5})t + (5.3 \times 10^{-5}) = 0$

$t = \dfrac{(1.3 \times 10^{-5}) \pm \sqrt{(1.3 \times 10^{-5})^2 - 4(4.2 \times 10^{-8})(5.3 \times 10^{-5})}}{2(4.2 \times 10^{-8})}$

$= \dfrac{1.3 \times 10^{-5} \pm \sqrt{1.60096 \times 10^{-10}}}{8.4 \times 10^{-8}} \approx \dfrac{1.3 \times 10^{-5} \pm 1.2653 \times 10^{-5}}{8.4 \times 10^{-8}}$

$t \approx 4$ or $t \approx 305$

Using the second derivative test:

$\dfrac{d^2S}{dt^2} = S_0[(-1.3 \times 10^{-5}) + (8.4 \times 10^{-8})t]$

For $t = 4$, $\dfrac{d^2S}{dt^2} = S_0[(-1.3 \times 10^{-5}) + (8.4 \times 10^{-8})(4)]$ (maximum)

For $t = 305$, $\dfrac{d^2S}{dt^2} = S_0[(-1.3 \times 10^{-5}) + (8.4 \times 10^{-8})(305)]$ (minimum)

The water has maximum density at $t = 4$.

Chapter 4 Review   (pages 298-290)

1. $\left(c, f(c)\right)$ 2. $\left(d, f(d)\right), \left(h, f(h)\right)$ 3. $a < x < d,\ h < x < b$

4. $d < x < h$ 5. $f(a)$ 6. $\left(e, f(e)\right)$

7a. $\dfrac{dy}{dx} = 4 - 3x^2$

Set $\dfrac{dy}{dx} = 0$:

$4 - 3x^2 = 0$

$x^2 = \dfrac{4}{3}$

$x = \pm\sqrt{\dfrac{4}{3}} = \dfrac{\pm 2}{\sqrt{3}} = \dfrac{\pm 2\sqrt{3}}{3}$

interval	$x < \dfrac{-2\sqrt{3}}{3}$	$\dfrac{-2\sqrt{3}}{3} < x < \dfrac{2\sqrt{3}}{3}$	$x > \dfrac{2\sqrt{3}}{3}$
sign of $\dfrac{dy}{dx}$	$-$	$+$	$-$
result: y is	decreasing	increasing	decreasing

7b. $\dfrac{d^2y}{dx^2} = -6x$

Set $\dfrac{d^2y}{dx^2} = 0$: $\quad -6x = 0$

$x = 0$

interval	$x < 0$	$x > 0$
sign of $\dfrac{d^2y}{dx^2}$	$+$	$-$
result: y is	concave up	concave down

point of inflection $(0,0)$

7c. *local minimum* $\left(\dfrac{-2\sqrt{3}}{3}, \dfrac{-16\sqrt{3}}{9}\right)$

local maximum $\left(\dfrac{2\sqrt{3}}{3}, \dfrac{16\sqrt{3}}{9}\right)$

7d. none

7e. *y-intercept:* $(0,0)$ *x-intercept:* $4x - x^3 = 0$ *symmetry:* origin

$x(4 - x^2) = 0$

$x = 0 \quad x = 2 \quad x = -2$

asymptotes: none

$(0,0) \quad (2,0) \quad (-2,0)$

7f.

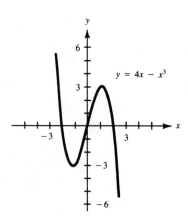

$y = 4x - x^3$

8a. $f'(x) = 8x - 3x^2$

Set $f'(x) = 0$:

$8x - 3x^2 = 0$

$x(8 - 3x) = 0$

$x = 0 \quad x = \frac{8}{3}$

interval	$x < 0$	$0 < x < \frac{8}{3}$	$x > \frac{8}{3}$
sign of $f'(x)$	$-$	$+$	$-$
result: $f(x)$ is	decreasing	increasing	decreasing

8b. $f''(x) = 8 - 6x$

Set $f''(x) = 0$:

$8 - 6x = 0$

$x = \frac{4}{3}$

interval	$x < \frac{4}{3}$	$x > \frac{4}{3}$
sign of $f''(x)$	$+$	$-$
result: $f(x)$ is	concave up	concave down

point of inflection: $\left(\frac{4}{3}, \frac{128}{27}\right)$

8c. local minimum: $(0,0)$ local maximum: $\left(\frac{8}{3}, \frac{256}{27}\right)$

8d. none

8e. y-intercept: $(0,0)$ x-intercept: $4x^2 - x^3 = 0$ symmetry: none

$x^2(4 - x) = 0$

$x = 0 \quad x = 4$

asymptotes: none $(0,0) \quad (4,0)$

8f.

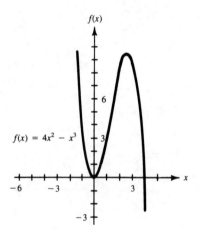

$f(x) = 4x^2 - x^3$

9a. $f'(x) = 4x^3 - 20x$

Set $f'(x) = 0$: $\quad 4x^3 - 20x = 0$

$$4x(x^2 - 5) = 0$$
$$x = 0 \quad x = \pm\sqrt{5}$$

interval	$x < -\sqrt{5}$	$-\sqrt{5} < x < 0$	$0 < x < \sqrt{5}$	$x > \sqrt{5}$
sign of $f'(x)$	$-$	$+$	$-$	$+$
result: $f(x)$ is	decreasing	increasing	decreasing	increasing

9b. $f''(x) = 12x^2 - 20$

Set $f''(x) = 0$: $\quad 12x^2 - 20 = 0$

$$4(3x^2 - 5) = 0$$
$$x^2 = \frac{5}{3}$$
$$x = \pm\sqrt{\frac{5}{3}} = \pm\frac{\sqrt{15}}{3}$$

interval	$x < \frac{-\sqrt{15}}{3}$	$\frac{-\sqrt{15}}{3} < x < \frac{\sqrt{15}}{3}$	$x > \frac{\sqrt{15}}{3}$
sign of $f''(x)$	$+$	$-$	$+$
result: $f(x)$ is	concave up	concave down	concave up

points of inflection: $\left(\dfrac{-\sqrt{15}}{3}, \dfrac{-44}{9}\right)\left(\dfrac{\sqrt{15}}{3}, \dfrac{-44}{9}\right)$

9c. *local minima:* $(-\sqrt{5}, -16)\ (\sqrt{5}, -16)$ $\qquad$ *local maximum:* $(0, 9)$

9d. *absolute minimum:* $f(-\sqrt{5}) = f(\sqrt{5}) = -16$

9e. *y-intercept:* $\quad f(0) = 9 \quad$ *x-intercept:* $\quad x^4 - 10x^2 + 9 = 0 \quad$ *symmetry:* y-axis

$\qquad\qquad\qquad (0, 9) \qquad\qquad\qquad\qquad (x^2 - 1)(x^2 - 9) = 0$

$\qquad\qquad\qquad\qquad\qquad\qquad\qquad\qquad (x + 1)(x - 1)(x + 3)(x - 3)$

$\qquad\qquad\qquad\qquad\qquad\qquad\qquad\qquad x = -1 \quad x = 1 \quad x = -3 \quad x = 3$

asymptotes: none $\qquad\qquad\qquad\qquad (-1, 0) \quad (1, 0) \quad (-3, 0) \quad (3, 0)$

9f.

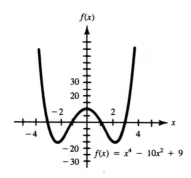

$f(x) = x^4 - 10x^2 + 9$

10a. $\dfrac{dy}{dx} = 4x^3 - 10x$

Set $\dfrac{dy}{dx} = 0$:

$$4x^3 - 10x = 0$$
$$2x(2x^2 - 5) = 0$$
$$x = 0 \qquad x = \pm\sqrt{\dfrac{5}{2}} = \pm\dfrac{\sqrt{10}}{2}$$

interval	$-2 < x < \dfrac{-\sqrt{10}}{2}$	$\dfrac{-\sqrt{10}}{2} < x < 0$	$0 < x < \dfrac{\sqrt{10}}{2}$	$\dfrac{\sqrt{10}}{2} < x < 3$
sign of $\dfrac{dy}{dx}$	$-$	$+$	$-$	$+$
result: y is	decreasing	increasing	decreasing	increasing

10b. $\dfrac{d^2y}{dx^2} = 12x^2 - 10$

Set $\dfrac{d^2y}{dx^2} = 0$:

$$12x^2 - 10 = 0$$
$$2(6x^2 - 5) = 0$$
$$x^2 = \dfrac{5}{6}$$
$$x = \pm\sqrt{\dfrac{5}{6}} = \pm\dfrac{\sqrt{30}}{6}$$

interval	$-2 < x < \dfrac{-\sqrt{30}}{6}$	$\dfrac{-\sqrt{30}}{6} < x < \dfrac{\sqrt{30}}{6}$	$\dfrac{\sqrt{30}}{6} < x < 3$
sign of $\dfrac{d^2y}{dx^2}$	$+$	$-$	$+$
result: y is	concave up	concave down	concave up

points of inflection: $\left(\dfrac{-\sqrt{30}}{6}, \dfrac{19}{36}\right)\left(\dfrac{\sqrt{30}}{6}, \dfrac{19}{36}\right)$

10c. *local minima:* $\left(\dfrac{-\sqrt{10}}{2}, \dfrac{-9}{4}\right)\left(\dfrac{\sqrt{10}}{2}, \dfrac{-9}{4}\right)$ *local maximum:* $(0,4)$

10d. *absolute minimum:* $f\left(\dfrac{-\sqrt{10}}{2}\right) = f\left(\dfrac{\sqrt{10}}{2}\right) = \dfrac{-9}{4}$

absolute maximum: $f(3) = 40$

10e. *y-intercept:* $y = 4$ *x-intercepts:* $x^4 - 5x^2 + 4 = 0$
$(0,4)$ $(x^2 - 1)(x^2 - 4) = 0$
                                        $(x + 1)(x - 1)(x + 2)(x - 2)$

symmetry: none $x = -1 \quad x = 1 \quad x = -2 \quad x = 2$
(since $-2 \le x \le 3$) $(-1,0) \quad (1,0) \quad (-2,0) \quad (2,0)$

asymptotes: none

10f.

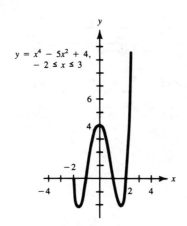

$y = x^4 - 5x^2 + 4,$
$-2 \le x \le 3$

11a. $y = (x-1)^{1/3}$

$\dfrac{dy}{dx} = \dfrac{1}{3}(x-1)^{-2/3}$

$= \dfrac{1}{3\sqrt[3]{(x-1)^2}}$

$\dfrac{dy}{dx}$ is undefined for $x = 1$

interval	$x < 1$	$x > 1$
sign of $\dfrac{dy}{dx}$	$+$	$+$
result: y is	increasing	increasing

11b. $\dfrac{d^2y}{dx^2} = \dfrac{1}{3}\left(\dfrac{-2}{3}\right)(x-1)^{-5/3}$

$= \dfrac{-2}{9\sqrt[3]{(x-1)^5}}$

$\dfrac{d^2y}{dx^2}$ is undefined for $x = 1$

point of inflection: $(1, 0)$

interval	$x < 1$	$x > 1$
sign of $\dfrac{d^2y}{dx^2}$	$+$	$-$
result: y is	concave up	concave down

11c. none

11d. none

11e. y-intercept: $y = \sqrt[3]{0-1}$
$= -1$
$(0, -1)$

asymptotes: none

x-intercept: $\sqrt[3]{x-1} = 0$
$x - 1 = 0$
$x = 1$
$(1, 0)$

symmetry: none

11f.

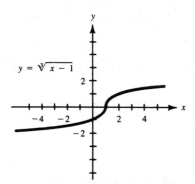

$y = \sqrt[3]{x-1}$

12a. $f'(x) = \frac{2}{3}x^{-1/3} = \frac{2}{3\sqrt[3]{x}}$

$f'(x)$ is undefined for $x = 0$

interval	$x < 0$	$x > 0$
sign of $f'(x)$	−	+
result: $f(x)$ is	decreasing	increasing

12b. $f''(x) = \frac{2}{3}\left(\frac{-1}{3}\right)x^{-4/3} = \frac{-2}{9\sqrt[3]{x^4}}$

$f''(x)$ is undefined for $x = 0$

point of inflection: none

interval	$x < 0$	$x > 0$
sign of $f''(x)$	−	−
result: $f(x)$ is	concave down	concave down

12c. local minimum: $(0,0)$

12d. absolute minimum: $f(0) = 0$

12e. y-intercept: $f(0) = 0$ x-intercept: $x^{2/3} = 0$ symmetry: y-axis
              $(0,0)$ $(0,0)$

asymptotes: none

12f.

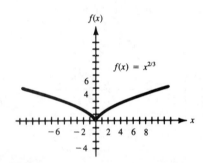

$f(x) = x^{2/3}$

13a. $\frac{dy}{dx} = 5x^4 - 12x^2$

Set $\frac{dy}{dx} = 0$:

$$5x^4 - 12x^2 = 0$$
$$x^2(5x^2 - 12) = 0$$
$$x^2 = 0 \quad x^2 = \frac{12}{5}$$
$$x = 0 \quad x = \pm\sqrt{\frac{12}{5}} = \pm\frac{2\sqrt{15}}{5}$$

interval	$x < \frac{-2\sqrt{15}}{5}$	$\frac{-2\sqrt{15}}{5} < x < 0$	$0 < x < \frac{2\sqrt{15}}{5}$	$x > \frac{2\sqrt{15}}{5}$
sign of $\frac{dy}{dx}$	+	−	−	+
result: y is	increasing	decreasing	decreasing	increasing

13b. $\dfrac{d^2y}{dx^2} = 20x^3 - 24x$

Set $\dfrac{d^2y}{dx^2} = 0$:

$$20x^3 - 24x = 0$$
$$4x(5x^2 - 6) = 0$$
$$4x = 0 \quad x^2 = \frac{6}{5}$$
$$x = 0 \quad x = \pm\sqrt{\frac{6}{5}} = \pm\frac{\sqrt{30}}{5}$$

interval	$x < \frac{-\sqrt{30}}{5}$	$\frac{-\sqrt{30}}{5} < x < 0$	$0 < x < \frac{\sqrt{30}}{5}$	$x > \frac{\sqrt{30}}{5}$
sign of $\frac{d^2y}{dx^2}$	$-$	$+$	$-$	$+$
result: y is	concave down	concave up	concave down	concave up

points of inflection: $\left(\dfrac{-\sqrt{30}}{5}, \dfrac{84\sqrt{30}}{125}\right)$, $(0,0)$, $\left(\dfrac{\sqrt{30}}{5}, \dfrac{-84\sqrt{30}}{125}\right)$

13c. *local minimum:* $\left(\dfrac{2\sqrt{15}}{5}, \dfrac{-192\sqrt{15}}{125}\right)$ *local maximum:* $\left(\dfrac{-2\sqrt{15}}{5}, \dfrac{192\sqrt{15}}{125}\right)$

13d. none

13e. *y-intercept:* $y = 0$ *x-intercepts:* $x^5 - 4x^3 = 0$ *symmetry:* origin

$(0,0)$

$$x^3(x^2 - 4) = 0$$
$$x^3(x + 2)(x - 2) = 0$$
$$x = 0 \quad x = -2 \quad x = 2$$
$$(0,0) \quad (-2,0) \quad (2,0)$$

asymptotes: none

13f.

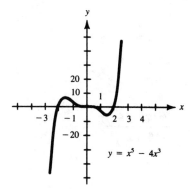

$y = x^5 - 4x^3$

14a. $y = 4x^{-2} - x$

$\dfrac{dy}{dx} = -8x^{-3} - 1$

$= \dfrac{-8}{x^3} - 1$

Set $\dfrac{dy}{dx} = 0$: $\dfrac{-8}{x^3} - 1 = 0$

$$-8 = x^3$$
$$-2 = x \text{ (not in domain)}$$

$\dfrac{dy}{dx} < 0$ for $x > 0$ (decreasing)

14b. $\dfrac{d^2y}{dx^2} = 24x^{-4} = \dfrac{24}{x^4}$

$\dfrac{d^2y}{dx^2} > 0$ for $x > 0$

(concave up)

14c. none **14d.** none

14e. *y-intercept*: none   *x-intercept*: $\dfrac{4}{x^2} - x = 0$ *symmetry*: none

$$4 = x^3$$
$$\sqrt[3]{4} = x$$

vertical asymptote: none $(\sqrt[3]{4}, 0)$

14f.

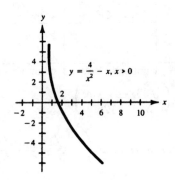

$y = \dfrac{4}{x^2} - x, \ x > 0$

15a. $\dfrac{dy}{dx} = \dfrac{(16 - x^2)(1) - x(-2x)}{(16 - x^2)^2} = \dfrac{x^2 + 16}{(16 - x^2)^2}$

There are no critical values. However, y is undefined for $x = \pm 4$.

interval	$x < -4$	$-4 < x < 4$	$x > 4$
sign of $\dfrac{dy}{dx}$	$+$	$+$	$+$
result: y is	increasing	increasing	increasing

15b. $\dfrac{d^2y}{dx^2} = \dfrac{(16 - x^2)^2(2x) - (x^2 + 16)(2)(16 - x^2)(-2x)}{\left[(16 - x^2)^2\right]^2}$

$$= \dfrac{2x(16 - x^2)[(16 - x^2) - (x^2 + 16)(-2)]}{(16 - x^2)^4} = \dfrac{2x(x^2 + 48)}{(16 - x^2)^3}$$

$\dfrac{d^2y}{dx^2} = 0$ for $x = 0$. Also, y is undefined for $x = \pm 4$.

interval	$x < -4$	$-4 < x < 0$	$0 < x < 4$	$x > 4$
sign of $\dfrac{d^2y}{dx^2}$	$+$	$-$	$+$	$-$
result: y is	concave up	concave down	concave up	concave down

point of inflection $(0, 0)$

15c. none   **15d.** none

15e.   *y-intercept:* $y = \dfrac{0}{16 - 0}$ *x-intercept:* $\dfrac{x}{16 - x^2} = 0$ *symmetry:* origin

$(0,0)$ $x = 0$

$(0,0)$

vertical asymptotes: $x = 4,\ x = -4$

horizontal asymptote: $y = 0$

15f.

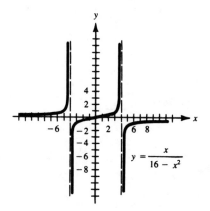

$y = \dfrac{x}{16 - x^2}$

16a. $y = \dfrac{x}{(1 - x^2)^{1/2}}$ $(-1 < x < 1)$

$\dfrac{dy}{dx} = \dfrac{(1 - x^2)^{1/2}(1) - x\left(\frac{1}{2}\right)(1 - x^2)^{-1/2}(-2x)}{\left[(1 - x^2)^{1/2}\right]^2} = \dfrac{(1 - x^2)^{-1/2}[(1 - x^2) + x^2]}{(1 - x^2)}$

$= \dfrac{1}{(1 - x^2)^{3/2}} = (1 - x^2)^{-3/2}$ There are no critical values.

interval	$-1 < x < 1$
sign of $\dfrac{dy}{dx}$	$+$
result: y is	increasing

16b. $\dfrac{d^2y}{dx^2} = \dfrac{-3}{2}(1 - x^2)^{-5/2}(-2x)$

$= \dfrac{3x}{\sqrt{(1 - x^2)^5}}$

$\dfrac{d^2y}{dx^2} = 0$ for $x = 0$.

point of inflection: $(0,0)$

interval	$-1 < x < 0$	$0 < x < 1$
sign of $\dfrac{d^2y}{dx^2}$	$-$	$+$
result: y is	concave down	concave up

16c.   none   **16d.**   none

16e. *y-intercept:* $y = \dfrac{0}{\sqrt{1-0}}$ *x-intercept:* $\dfrac{x}{\sqrt{1-x^2}} = 0$ *symmetry:* origin

        $(0,0)$ $x = 0$

                                      $(0,0)$

     vertical asymptotes: $x = -1, \ x = 1$

     horizontal asymptote: none

16f.

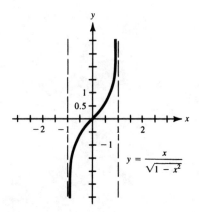

17a.   Let $x =$ width of the yard

       Let $y =$ length of the yard   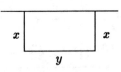

$$2x + y = 80$$
$$y = 80 - 2x$$

Maximize the area: Using the second derivative test:

$$A = xy \qquad\qquad\qquad\qquad\qquad \frac{d^2A}{dx^2} = -4 \ \text{(maximum)}$$

$$A = x(80 - 2x) \qquad\qquad\qquad\quad y = 80 - 2(20) = 40$$

$$A = 80x - 2x^2$$

$$\frac{dA}{dx} = 80 - 4x$$

Set $\dfrac{dA}{dx} = 0$: $80 - 4x = 0$

                                $x = 20$

18. $x + 3y = 72, \ x = 72 - 3y$

Maximize the product: Using the second derivative test:

$$P = xy \qquad\qquad\qquad\qquad\qquad \frac{d^2P}{dy^2} = -6 \ \text{(maximum)}$$
$$P = (72 - 3y)y = 72y - 3y^2$$

$$\frac{dP}{dy} = 72 - 6y \qquad\qquad\qquad\quad x = 72 - 3(12) = 36$$

Set $\dfrac{dP}{dy} = 0$: $72 - 6y = 0$

                                $y = 12$

19a. $R(x) = xp = x\left(\dfrac{40}{\sqrt{x}}\right) = \dfrac{40x}{\sqrt{x}} \cdot \dfrac{\sqrt{x}}{\sqrt{x}} = \dfrac{40x\sqrt{x}}{x} = 40\sqrt{x}$

19b. $P(x) = R(x) - C(x) = 40\sqrt{x} - (0.5x + 400) = 40x^{1/2} - 0.5x - 400$

$P'(x) = 40\left(\dfrac{1}{2}\right)x^{-1/2} - 0.5$

$\qquad = \dfrac{20}{\sqrt{x}} - 0.5$

Set $P'(x) = 0$: $\qquad \dfrac{20}{\sqrt{x}} - 0.5 = 0$

$\qquad\qquad\qquad\qquad\qquad 20 = 0.5\sqrt{x}$

$\qquad\qquad\qquad\qquad\qquad 40 = \sqrt{x}$

$\qquad\qquad\qquad\qquad\qquad 1600 = x$

Using the second derivative test:

$P''(x) = 20\left(\dfrac{-1}{2}\right)x^{-3/2} = \dfrac{-10}{\sqrt{x^3}}$

$P''(1600) = \dfrac{-10}{64000}$ (maximum)

19c. $P(1600) = 40\sqrt{1600} - 0.5(1600) - 400 = \$400$

19d. $P = \dfrac{40}{\sqrt{1600}} = \$1$

20. $\dfrac{dR}{dx} = \dfrac{1}{16}(63x^2 - 254x + 154)$

Set $\dfrac{dR}{dx} = 0$: $\qquad \dfrac{1}{16}(63x^2 - 254x + 154) = 0$

$\qquad\qquad\qquad\qquad\qquad 63x^2 - 254x + 154 = 0$

$x = \dfrac{254 \pm \sqrt{(254)^2 - 4(63)(154)}}{2(63)} \approx \dfrac{254 \pm 160.34}{126}$

$x \approx 0.74 \qquad x \approx 3.29$ (not in domain)

When $x = 0$, $R = \dfrac{1}{16}(80) = 5$

When $x = 0.74 \qquad R \approx \dfrac{1}{16}(132.92) \approx 8.31$ (maximum)

When $x = 2 \qquad R = \dfrac{1}{16}(48) = 3$ (minimum)

Radiation is lowest for $x = 2$ and highest for $x = 0.74$

Chapter 4 Test (pages 300-301)

1. b (Consider $f(x) = x^3$)

2. b (Consider $f(x) = x^4$)

3. b (Consider $f(x) = x^{2/3}$)

4. b (Consider $f(x) = -x^2$, $-1 \le x \le 2$)

5. $f'(x) = 4x^3 - 12x$

   $f''(x) = 12x^2 - 12$

   Set $f''(x) = 0$: $12x^2 - 12 = 0$

   $12(x^2 - 1) = 0$

   $x = \pm 1$

interval	$x < -1$	$-1 < x < 1$	$x > 1$
sign of $f''(x)$	$+$	$-$	$+$
result: $f(x)$ is	concave up	concave down	concave up

   points of inflection: $(-1, -5)(1, -5)$ (c)

6.   c

7. $y = x^{1/2} + (x-1)^3$

   $\dfrac{dy}{dx} = \dfrac{1}{2}x^{-1/2} + 3(x-1)^2$

   $= \dfrac{1}{2\sqrt{x}} + 3(x-1)^2$ (a)

8.   b   (The function is concave down at a critical value.)

9.   *critical values:*

   $\dfrac{dy}{dx} = -6x^2 + 10x + 4$

   Set $\dfrac{dy}{dx} = 0$: $-6x^2 + 10x + 4 = 0$

   $-2(3x^2 - 5x - 2) = 0$

   $-2(3x + 1)(x - 2) = 0$

   $x = \dfrac{-1}{3}$ $x = 2$

interval	$x < -\frac{1}{3}$	$-\frac{1}{3} < x < 2$	$x > 2$
sign of $\dfrac{dy}{dx}$	$-$	$+$	$-$
result: y is	decreasing	increasing	decreasing

   local minimum $\left(-\frac{1}{3}, \frac{-46}{27}\right)$ *local maximum* $(2, 11)$

   concavity: $\dfrac{d^2y}{dx^2} = -12x + 10$

interval	$x < \frac{5}{6}$	$x > \frac{5}{6}$
sign of $\dfrac{d^2y}{dx^2}$	$+$	$-$
result: y is	concave up	concave down

   Set $\dfrac{d^2y}{dx^2} = 0$: $-12x + 10 = 0$

   $x = \dfrac{5}{6}$

   point of inflection: $\left(\frac{5}{6}, \frac{251}{54}\right)$

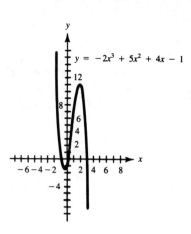

$y = -2x^3 + 5x^2 + 4x - 1$

10a,b. $f'(x) = 3x^2 - 10x - 8$

Set $f'(x) = 0$: $3x^2 - 10x - 8 = 0$

$$(3x + 2)(x - 4) = 0$$

$$x = \frac{-2}{3} \qquad x = 4$$

interval	$x < \frac{-2}{3}$	$\frac{-2}{3} < x < 4$	$x > 4$
sign of $f'(x)$	$+$	$-$	$+$
result: $f(x)$ is	increasing	decreasing	increasing

10c,d. $f''(x) = 6x - 10$

Set $f''(x) = 0$: $6x - 10 = 0$

$$x = \frac{5}{3}$$

interval	$x < \frac{5}{3}$	$x > \frac{5}{3}$
sign of $f''(x)$	$-$	$+$
result: $f(x)$ is	concave down	concave up

10e. $\left(\frac{5}{3}, \frac{-610}{27}\right)$

11. $P'(x) = \dfrac{(x^2 + 16)(8) - 8x(2x)}{(x^2 + 16)^2} + 0 = \dfrac{128 - 8x^2}{(x^2 + 16)^2} = \dfrac{-8(x^2 - 16)}{(x^2 + 16)^2}$

Set $P'(x) = 0$: $\dfrac{-8(x^2 - 16)}{(x^2 + 16)^2} = 0$

$$-8(x^2 - 16) = 0$$

$$x^2 - 16 = 0$$

$x = 4$ since x cannot be negative

interval	$0 < x < 4$	$x > 4$
sign of $P'(x)$	$+$	$-$
result: $P(x)$ is	increasing	decreasing

The profit is a maximum when $x = 4$. ($4000)

12a. $R(x) = px = (75 - \frac{x}{4})x = 75x - \frac{x^2}{4}$

12b,c. $R'(x) = 75 - \frac{x}{2}$

Set $R'(x) = 0$: $75 - \frac{x}{2} = 0$

$\frac{-x}{2} = -75$

$x = 150$

interval	$0 < x < 150$	$150 < x < 260$
sign of $R'(x)$	+	−
result: $R(x)$ is	increasing	decreasing

12d. The maximum occurs when $x = 150$.

$p = 75 - \frac{150}{4} = \$37.50$

13a. $P'(s) = 12000 - 200s$
$P'(40) = 12000 - 200(40) = 4000$
so the profit is increasing

13b. Set $P'(s) = 0$:
$12000 - 200s = 0$
$s = 60$

interval	$0 < s < 60$	$s > 60$
sign of $P'(x)$	+	−
result: $P(x)$ is	increasing	decreasing

Increasing for $0 < s < 60$.

13c. $60

14. $R(x) = 425 - 2500(x+6)^{-1} - 4x$
$R'(x) = 2500(x+6)^{-2} - 4$
$= \frac{2500}{(x+6)^2} - 4$

interval	$0 < x < 19$	$x > 19$
sign of $R'(x)$	+	−
result: $R(x)$ is	increasing	decreasing

Set $R'(x) = 0$: $\frac{2500}{(x+6)^2} - 4 = 0$

$2500 = 4(x+6)^2$

$625 = (x+6)^2$

$x + 6 = 25$ (since x is not negative)

$x = 19$

$R(19) = 425 - \frac{2500}{19+6} - 4(19) = \$249$

15.
$$2x + 2y = 160$$
$$2y = 160 - 2x$$
$$y = 80 - x$$

Maximize the area:

$$A = xy$$
$$A = x(80 - x)$$
$$A = 80x - x^2$$
$$\frac{dA}{dx} = 80 - 2x$$

Set $\frac{dA}{dx} = 0$: $80 - 2x = 0$
$$x = 40$$

Using the second derivative test:

$$\frac{d^2A}{dx^2} = -2 \text{ (maximum)}$$
$$y = 80 - 40 = 40$$
$$A = (40)(40) = 1600 \text{ in.}^2$$

CHAPTER 5: EXPONENTIAL AND LOGARITHMIC FUNCTIONS

Section 5.1   (pages 310-313)

1.   exponential growth

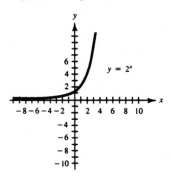

3.   exponential decay

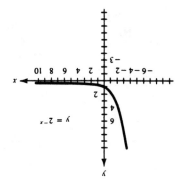

5.   exponential decay

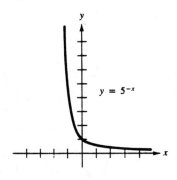

7.   neither

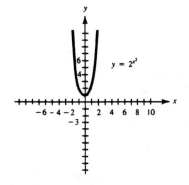

9. exponential growth

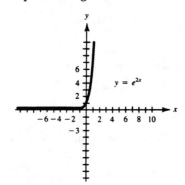

11. exponential growth

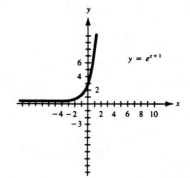

13. exponential growth

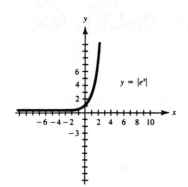

15. $2^x = 2^5$

$x = 5$

17. $10^x = \dfrac{1}{10^2}$

$10^x = 10^{-2}$

$x = -2$

19. $(9^2)^x = 9^{x+1}$

$9^{2x} = 9^{x+1}$

$2x = x + 1$

$x = 1$

21. $3^{x^2 + 2x} = 3^{-1}$

$x^2 + 2x = -1$

$x^2 + 2x + 1 = 0$

$(x+1)^2 = 0$

$x = -1$

23. $2^x = 16$

$2^x = 2^4$

$x = 4$

25a.

t	$(1.06)^t$
10	1.791
11	1.898
12	2.01

t is about 12 years.

25b.

t	$(1.12)^t$
5	1.762
6	1.974
7	2.211

t is a little more than 6 years.

25c.

t	$(1.10)^t$
3	1.331
4	1.464
5	1.610

$(1.10)^t = 1.50$ (150%) when t is between 4 and 5 years

27.　$A = 2000\left(1 + \dfrac{.10}{365}\right)^{(365)(2)} \approx \$2442.74$

29a.　$A = 2000e^{(.10)(2)} \approx \$2442.81$

29b.　The amount is higher than the answer from #27 ($2442.74) but lower than the answer from #28 ($2449.53).

31a.　$A = 1\left(1 + \dfrac{.10}{360}\right)^{(365)(1)} \approx \$1.10669 \approx \$1.11$

31b.　$0.10669 = 10.669\%$

31c.　$A = 1\left(1 + \dfrac{.10}{365}\right)^{(365)(1)} \approx 1.10515$

　　　The effective rate $(0.10515 = 10.515\%)$ is slightly lower than the answer in part b.

33.　$V = 100,000(1.08)^3 \approx \$125,971$

35a.

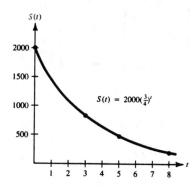

35b.　$S(0) = 2000\left(\dfrac{3}{4}\right)^0 = \$2000$

35c.　$S(3) = 2000\left(\dfrac{3}{4}\right)^3 = \$843.75$

37.　$3 = r(4)^{-1.2}$　　　$r = \dfrac{3}{4^{-1.2}} \approx 15.83$ roentgens

39a.　$d = 21.9\left[2^{(20-19)/12}\right] = 21.9\left[2^{1/12}\right] \approx 23.2$ cm

39b.　$d = 21.9\left[2^{(20-0)/12}\right] = 21.9\left[2^{20/12}\right] \approx 69.5$ cm

Section 5.1A (page 321)

1.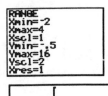

3.

5.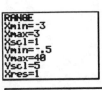

7.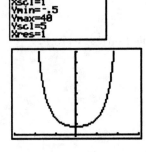

9.

11. $x = 0$, $x \approx -0.57$

13. $x = 0$, $x \approx -0.5$

15. $x = 0$

17. $x = 2$, $x \approx -1.69$

19. $x = 4$

21. $y = e^x$ for $x > 0$

    $y = e^{-x}$ for $x < 0$

    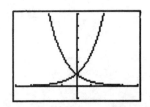

23. 2.7160201

25. 2.7182805

Section 5.2   (pages 329-332)

1. $\log_4 16 = 2$

3. $\log_5 \frac{1}{125} = -3$

5. $\log_{10}(0.01) = -2$

7. $3^2 = 9$

9. $10^2 = 100$

11. $e^{-1} = \frac{1}{e}$

13. $4^2 = 1 - x$
    $16 = 1 - x$
    $x = -15$

15. $3^{-1} = \frac{1}{x}$
    $\frac{1}{3} = \frac{1}{x}$
    $3 = x$

17. $x^2 = 9$
    $x = 3$ (since $x > 0$)

19.

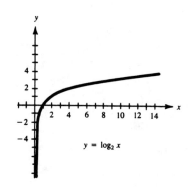
$y = \log_2 x$

21.

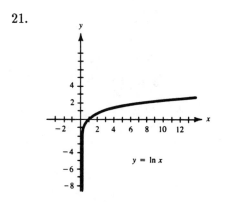
$y = \ln x$

23.

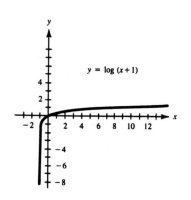
$y = \log (x + 1)$

25. 6.7940

27. 5.4600

29. $\log_b x + \log_b y^2$

$= \log_b x + 2 \log_b y$

31. $\log_b x - \log_b (y + z)$

33. $\log_b (x + y)^{1/3}$

$= \frac{1}{3} \log_b (x + y)$

35. $\log_5 0.012 = \dfrac{\log 0.012}{\log 5}$

≈ -2.7480704

37. $\log 15^{2x+1} = \log 85$

$(2x + 1)\log 15 = \log 85$

$2x + 1 = \dfrac{\log 85}{\log 15}$

$2x = \dfrac{\log 85}{\log 15} - 1$

$x = \dfrac{\log 85}{2 \log 15} - \dfrac{1}{2}$

$x \approx 0.3203$

39. $\log 10^{2x-3} = \log 5$

$(2x - 3)\log 10 = \log 5$

$(2x - 3)(1) = \log 5$

$2x = \log 5 + 3$

$x = \dfrac{\log 5}{2} + \dfrac{3}{2}$

$x \approx 1.8495$

41. $\ln e^{x^2} = \ln 10$

$x^2 \ln e = \ln 10$

$x^2 = \ln 10$

$x = \pm \sqrt{\ln 10}$

$x \approx \pm 1.5174$

43. $\left(\dfrac{1}{2}\right)^{3x-1} = 17$

$\log \left(\dfrac{1}{2}\right)^{3x-1} = \log 17$

$(3x - 1)\log \left(\dfrac{1}{2}\right) = \log 17$

$3x - 1 = \dfrac{\log 17}{\log \left(\frac{1}{2}\right)}$

$3x = \dfrac{\log 17}{\log \left(\frac{1}{2}\right)} + 1$

$x = \dfrac{\log 17}{3 \log \left(\frac{1}{2}\right)} + \dfrac{1}{3}$

$x \approx -1.0292$

45a.

$g(1) = e^1 = e$

$f[g(1)] = f(e) = \ln e = 1$

$g(10) = e^{10}$

$f[g(10)] = f(e^{10}) \quad = \ln e^{10}$

$= 10 \ln e = 10$

$g(100) = e^{100}$

$f[g(100)] = f(e^{100}) \quad = \ln e^{100}$

$= 100 \ln e = 100$

$g\left(\frac{1}{10}\right) = e^{1/10}$

$f\left[g\left(\frac{1}{10}\right)\right] = f(e^{1/10}) \quad = \ln e^{1/10}$

$= \frac{1}{10} \ln e = \frac{1}{10}$

$f(1) = \ln 1 = 0$

$g[f(1)] = g(0) = e^0 = 1$

$f(10) = \ln 10$

$g[f(10)] = g(\ln 10) = e^{\ln 10} = 10$

$f(100) = \ln 100$

$g[f(100)] = g(\ln 100) = e^{\ln 100} = 100$

$f\left(\frac{1}{10}\right) = \ln\left(\frac{1}{10}\right)$

$g\left[f\left(\frac{1}{10}\right)\right] = g\left[\ln\left(\frac{1}{10}\right)\right] = e^{\ln (1/10)} = \frac{1}{10}$

45b.

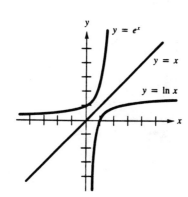

47a. $N(0) = 2500 + 250 \ln 1 = 2500$

47b. $N(8) = 2500 + 250 \ln (8 + 1) \approx 3049$

47c. $N(20) = 2500 + 250 \ln (20 + 1) \approx 3261$

49. $\quad 3000 = 2000(1.015)^{4t}$

$1.5 = (1.015)^{4t}$

$\log 1.5 = \log (1.015)^{4t}$

$\log 1.5 = 4t \log (1.015)$

$\dfrac{\log 1.5}{4 \log (1.015)} = t$

$t \approx 6.8 \text{ yr.}$

51a. $S(0) = 3000\left(\frac{1}{2}\right)^0 = \$3000$

51b. $3000\left(\frac{1}{2}\right)^t = 1500$

$\left(\frac{1}{2}\right)^t = \frac{1}{2}$

$t = 1 \text{ month}$

53a. $R(x) = px = \dfrac{300x}{\ln (x + 4)}$

53b. $R(20) = \dfrac{300(20)}{\ln (20 + 4)} \approx \$1887.95$

55a. $R = 70 \ln(0 + e) = 70$

55b. $R = 70 \ln (2 + e) \approx 109$

57a.
$$20,000 = \frac{(10,000)(50,000)}{10,000 + (50,000 - 10,000)e^{(-50,000)(K)(4)}}$$

$$20,000(10,000 + 40,000e^{-200,000K}) = (10,000)(50,000)$$

$$2(1 + 4e^{-200,000K}) = (1)(5)$$

$$2 + 8e^{-200,000K} = 5$$

$$8e^{-200,000K} = 3$$

$$e^{-200,000K} = \frac{3}{8}$$

$$-200,000K = \ln\frac{3}{8}$$

$$K = \frac{\ln\frac{3}{8}}{-200,000}$$

$$K \approx 4.9 \times 10^{-6}$$

57b.
$$N(t) = \frac{(10,000)(50,000)}{10,000 + 40,000e^{(-50,000)(4.9 \times 10^{-6}t)}} = \frac{5}{1 + 4e^{-0.245t}}$$

where $N(t)$ is measured in ten thousands

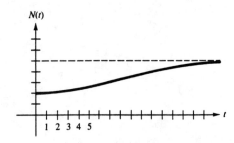

59a. $M = \log\left(\frac{1,000,000 A_0}{A_0}\right) = \log(1,000,000) = 6$

59b. $7.2 = \log\frac{A}{A_0}$

$$10^{7.2} = \frac{A}{A_0}$$

$$A = 10^{7.2}A_0 \approx (1.5849 \times 10^7)A_0$$

61a. $N = 0.25N_0$

$$0.25N_0 = N_0\left(\frac{1}{2}\right)^{x/5570}$$

$$0.25 = (0.5)^{x/5570}$$

$$\log(0.25) = \log(0.5)^{x/5570}$$

$$\log(0.25) = \frac{x}{5570}\log(0.5)$$

$$\frac{x}{5570} = \frac{\log(0.25)}{\log(0.5)}$$

$$x = (5570)\frac{\log(0.25)}{\log(0.5)}$$

$$x = 11,140 \text{ years}$$

61b. $N = 0.40N_0$

$$0.40N_0 = N_0\left(\frac{1}{2}\right)^{x/5570}$$

$0.40 = (0.50)^{x/5570}$

$\log (0.40) = \log (0.50)^{x/5570}$

$\log (0.40) = \frac{x}{5570} \log (0.50)$

$\frac{x}{5570} = \frac{\log (0.40)}{\log (0.50)}$

$x = (5570)\frac{\log (0.40)}{\log (0.50)}$

$x \approx 7363$

Section 5.2A   (page 338)

1.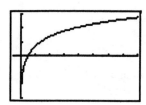

3.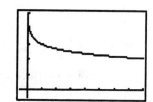

5. $y = \dfrac{\ln x}{\ln 7}$

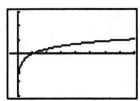

7.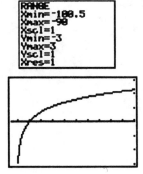

9. $x \approx 19.08$

11. $x = 1$

13a. The domain for $y = \ln x$ is $x > 0$.

For $y = \ln |x|, x \neq 0$.

13b. domain: $x > 0$

range: $y \geq 0$

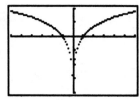

Domain: $x \neq 0$

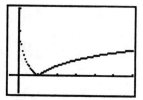

Domain: $x > 0$;   Range: $y \geq 0$

15.

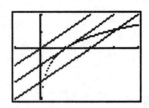

17a. domain: $x > 0$, $x \neq 1$ (since $\frac{1}{\ln 1} = \frac{1}{0}$)

17b. $\ln y = \ln x^{1/\ln x}$

$\ln y = \frac{1}{\ln x} \ln x$

$\ln y = 1$

$e^1 = y$ so $y = e$ (when $x > 0$, $x \neq 1$)

17c. $\frac{1}{\ln 1} = \frac{1}{0}$ There is a point on the calculator's graph when $x = 1$.

Section 5.3   (pages 344-346)

1. $\dfrac{dy}{dx} = \dfrac{1}{3x-1}(3) = \dfrac{3}{3x-1}$

3. $y = \ln x^{1/2} = \frac{1}{2}\ln x$

$\dfrac{dy}{dx} = \dfrac{1}{2} \cdot \dfrac{1}{x} = \dfrac{1}{2x}$

5. $\dfrac{dy}{dx} = 3(\ln x)^2 \cdot \frac{1}{x}$

$= \dfrac{3(\ln x)^2}{x}$

7. $y = \ln(2x+1) - \ln(x-1)$

$\dfrac{dy}{dx} = \dfrac{1}{2x+1}(2) - \dfrac{1}{x-1}(1)$

$= \dfrac{2}{2x+1} - \dfrac{1}{x-1}$

$= \dfrac{2(x-1) - 1(2x+1)}{(2x+1)(x-1)}$

$= \dfrac{-3}{(2x+1)(x-1)}$

9. $\dfrac{dy}{dx} = x^2 \cdot \frac{1}{x} + 2x \ln x$

$= x + 2x \ln x$

11. $\dfrac{dy}{dx} = \dfrac{\ln x(1) - x\left(\frac{1}{x}\right)}{(\ln x)^2}$

$= \dfrac{\ln x - 1}{(\ln x)^2}$

13. $\dfrac{dy}{dx} = \dfrac{(\ln 3x)\left(\frac{1}{2x} \cdot 2\right) - (\ln 2x)\left(\frac{1}{3x} \cdot 3\right)}{(\ln 3x)^2}$

$= \dfrac{\frac{\ln 3x}{x} - \frac{\ln 2x}{x}}{(\ln 3x)^2} = \dfrac{\ln 3x - \ln 2x}{x(\ln 3x)^2}$

15.
$$y = (\ln x)^{-2}$$
$$\frac{dy}{dx} = -2(\ln x)^{-3}\left(\frac{1}{x}\right)$$
$$= \frac{-2}{x(\ln x)^3}$$

17.
$$\frac{dy}{dx} = \frac{(x+1)\left(\frac{1}{x}\right) - (\ln x)(1)}{(x+1)^2}$$
$$= \frac{\frac{x+1}{x} - \ln x}{(x+1)^2}$$
$$= \frac{x+1-x \ln x}{x(x+1)^2}$$

19.
$$\frac{dy}{dx} = \frac{1}{\ln 10} \cdot \frac{1}{x}$$
$$= \frac{1}{x \ln 10}$$

21.
$$y = x \log_{10} x^{1/2}$$
$$y = \frac{1}{2}x \log_{10} x$$
$$\frac{dy}{dx} = \frac{1}{2}x\left(\frac{1}{\ln 10} \cdot \frac{1}{x}\right) + \frac{1}{2} \log_{10} x$$
$$= \frac{1}{2 \ln 10} + \frac{1}{2} \frac{\ln x}{\ln 10}$$
$$= \frac{1}{2 \ln 10}(1 + \ln x)$$

23.
$$\frac{dy}{dx} = \frac{1}{2x+1}(2)$$
$$= \frac{2}{2x+1}$$
$$= 2(2x+1)^{-1}$$
$$\frac{d^2y}{dx^2} = -2(2x+1)^{-2}(2)$$
$$= \frac{-4}{(2x+1)^2}$$

25. *critical value:*
$$y = \ln x^{1/2}$$
$$y = \frac{1}{2} \ln x$$
$$\frac{dy}{dx} = \frac{1}{2} \cdot \frac{1}{x} = \frac{1}{2x}$$
no critical values
$$\frac{dy}{dx} > 0 \text{ for } x > 0$$
(increasing)

concavity:
$$\frac{dy}{dx} = \frac{1}{2}x^{-1}$$
$$\frac{d^2y}{dx^2} = \frac{-1}{2}x^{-2} = \frac{-1}{2x^2}$$
$$\frac{d^2y}{dx^2} < 0 \text{ for } x > 0$$
(concave down)

x-intercept:
$$\ln \sqrt{x} = 0$$
$$e^0 = \sqrt{x}$$
$$1 = x$$
$$(1,0)$$

vertical asymptote: $\quad x = 0$

25. (continued)

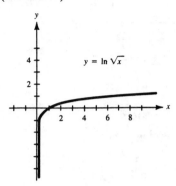

27.   *critical values:*

$$y = 2 \ln x$$

$$\frac{dy}{dx} = 2\left(\frac{1}{x}\right)$$

$$= \frac{2}{x}$$

interval	$x < 0$	$x > 0$
sign of $\dfrac{dy}{dx}$	$-$	$+$
result: y is	decreasing	increasing

There are no critical values but the function is not defined for $x = 0$.

concavity:

$$\frac{dy}{dx} = 2x^{-1}$$

$$\frac{d^2y}{dx^2} = -2x^{-2}$$

$$= \frac{-2}{x^2}$$

$$\frac{d^2y}{dx^2} < 0 \text{ (concave down)}$$

vertical asymptote: $x = 0$

x-intercepts: $\ln x^2 = 0$

$$e^0 = x^2$$

$$1 = x^2$$

$$\pm 1 = x$$

$$(1,0)\ (-1,0)$$

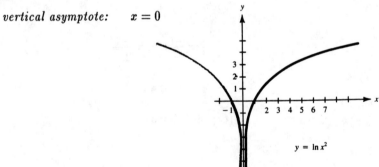

29.   *critical values:*

$$f'(x) = 2(\ln x)\left(\frac{1}{x}\right)$$

Set $f'(x) = 0$: $\dfrac{2 \ln x}{x} = 0$

$$2 \ln x = 0$$

$$\ln x = 0$$

$$x = e^0$$

$$x = 1$$

interval	$0 < x < 1$	$x > 1$
sign of $f'(x)$	$-$	$+$
result: $f(x)$ is	decreasing	increasing

local minimum: $(1,0)$

concavity:

$$f''(x) = \frac{x\left(\frac{2}{x}\right) - 2\ln x}{x^2}$$

$$= \frac{2 - 2\ln x}{x^2}$$

Set $f''(x) = 0$:

$$\frac{2 - 2\ln x}{x^2} = 0$$

$$2(1 - \ln x) = 0$$

$$\ln x = 1$$

$$x = e^1$$

interval	$0 < x < e$	$x > e$
sign of $f'(x)$	$+$	$-$
result: $f(x)$ is	concave up	concave down

point of inflection: $(e, 1)$

x-intercept: $(\ln x)^2 = 0$

$$\ln x = 0$$

$$x = e^0$$

$$x = 1$$

$$(1, 0)$$

vertical asymptote: $x = 0$

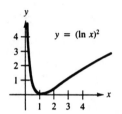

$y = (\ln x)^2$

31.   *critical values:*

$$\frac{dy}{dx} = \frac{1}{1-x}(-1)$$

$$= \frac{-1}{1-x}$$

There are no critical values.

$$\frac{dy}{dx} < 0 \text{ for } x < 1$$

(decreasing)

concavity:

$$\frac{dy}{dx} = -1(1-x)^{-1}$$

$$\frac{d^2y}{dx^2} = (1-x)^{-2}(-1)$$

$$= \frac{-1}{(1-x)^2}$$

$$\frac{d^2y}{dx^2} < 0 \text{ for } x < 1$$

(concave down)

x-intercept:

$$\ln(1-x) = 0$$

$$1 - x = e^0$$

$$1 - x = 1$$

$$x = 0$$

$$(0, 0)$$

vertical asymptote: $x = 1$

$y = \ln(1-x)$

33. $\dfrac{dP}{dx} = -10\left(\dfrac{1}{x}\right) = \dfrac{-10}{x}$

35a. $P(x) = 200 \ln (x + 1) - (4x + 6)$

$P'(x) = \dfrac{200}{x + 1} - 4$

Set $P'(x) = 0$: $\dfrac{200}{x + 1} - 4 = 0$

$$\dfrac{200}{x + 1} = 4$$

$$200 = 4x + 4$$

$$196 = 4x$$

$$49 = x$$

Using the second derivative test:

$$P'(x) = 200(x + 1)^{-1} - 4$$

$$P''(x) = -200(x + 1)^{-2}$$

$$P''(49) = \dfrac{-200}{(49 + 1)^2} \quad \text{(maximum)}$$

35b. $P(49) = 200 \ln (49 + 1) - [4(49) + 6]$

$$= \$580.40$$

37a. $S'(t) = \dfrac{60}{t + 1}$

$$S'(2) = \dfrac{60}{2 + 1} = 20$$

$$S'(4) = \dfrac{60}{4 + 1} = 12$$

37b. $S'(t) = 60(t + 1)^{-1}$

$$S''(t) = -60(t + 1)^{-2}$$

$$= \dfrac{-60}{(t + 1)^2}$$

$$S''(t) < 0 \quad \text{(decreasing rate)}$$

39. $N'(t) = t\left(\dfrac{1}{t + 1}\right) + (1) \ln(t + 1)$

$$= \dfrac{t}{t + 1} + \ln(t + 1)$$

$N'(30) = \dfrac{30}{30 + 1} + \ln(30 + 1)$

$$\approx 4.4017 \quad \text{(hundreds)}$$

Section 5.4   (pages 352-354)

1. $\dfrac{dy}{dx} = 3e^{3x}$

3. $\dfrac{dy}{dx} = 12 \cdot 6e^{6x}$

$$= 72e^{6x}$$

5. $\dfrac{dy}{dx} = -2xe^{1 - x^2}$

7. $\dfrac{dy}{dx} = xe^x + 1e^x$

$$= e^x(x + 1)$$

9. $\dfrac{dy}{dx} = \dfrac{(e^x + 1)(e^x) - e^x(e^x)}{(e^x + 1)^2}$

$$= \dfrac{e^x}{(e^x + 1)^2}$$

11. $\dfrac{dy}{dx} = x^2(-e^{-x}) + 2xe^{-x}$

$$= xe^{-x}(-x + 2)$$

13. $y = x$

$\dfrac{dy}{dx} = 1 \quad \left(\text{or } \dfrac{dy}{dx} = \dfrac{1}{e^x} \cdot e^x = 1\right)$

15. $y = x^2$

$$\dfrac{dy}{dx} = 2x$$

17. $y = x$

$\dfrac{dy}{dx} = 1$

19. $\dfrac{dy}{dx} = 4e^{4x}$

$\dfrac{d^2y}{dx^2} = 4 \cdot 4e^{4x}$

$\qquad = 16e^{4x}$

21. $\dfrac{dy}{dx} = 2(e^x + 1)e^x = 2e^{2x} + 2e^x$

$\dfrac{d^2y}{dx^2} = 2 \cdot 2e^{2x} + 2e^x = 4e^{2x} + 2e^x$

23. *critical values:*

$\dfrac{dy}{dx} = 5e^{5x-1}$

There are no critical values.

$\dfrac{dy}{dx}$ is always positive.

(increasing for all x)

concavity: $\dfrac{d^2y}{dx^2} = 25e^{5x-1}$

$\dfrac{d^2y}{dx^2}$ is always positive.

(concave up for all x)

horizontal asymptote: $y = 0$

y-intercept: $y = e^{5(0)-1}$

$\qquad\qquad y = e^{-1}$

$\qquad\qquad (0, \tfrac{1}{e})$

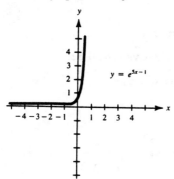

$y = e^{5x-1}$

25. *critical values:*

$\dfrac{dy}{dx} = xe^x + 1e^x$

$\qquad = e^x(x + 1)$

Set $\dfrac{dy}{dx} = 0$: $\quad e^x(x + 1) = 0$

$\qquad\qquad\qquad x + 1 = 0$

$\qquad\qquad\qquad x = -1$

interval	$x < -1$	$x > -1$
sign of $\dfrac{dy}{dx}$	$-$	$+$
result: y is	decreasing	increasing

local minimum: $(-1, \tfrac{-1}{e})$

concavity:

$$\frac{d^2y}{dx^2} = e^x(1) + e^x(x+1)$$

$$= e^x(x+2)$$

Set $\dfrac{d^2y}{dx^2} = 0$: $\quad e^x(x+2) = 0$

$$x + 2 = 0$$

$$x = -2$$

interval	$x < -2$	$x > -2$
sign of $\dfrac{d^2y}{dx^2}$	$-$	$+$
result: y is	concave down	concave up

point of inflection: $\quad \left(-2, \dfrac{-2}{e^2}\right)$

horizontal asymptote: $\quad y = 0$

y-intercept: $\quad y = 0e^0 = 0$

$(0,0)$

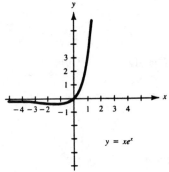

$y = xe^x$

27. *critical values:*

$$\frac{dy}{dx} = -e^x$$

There are no critical values.

$\dfrac{dy}{dx}$ is negative for all x.

(decreasing)

concavity: $\quad \dfrac{d^2y}{dx^2} = -e^x$

$\dfrac{d^2y}{dx^2}$ is negative for all x.

(concave down)

horizontal asymptote: $\quad y = 4$

y-intercept: $\quad y = 4 - e^0$

$$y = 3$$

$$(0,3)$$

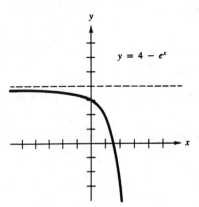

$y = 4 - e^x$

29. $\dfrac{dy}{dx} = \dfrac{(x+1)(3^x \ln 3) - (1)(3^x)}{(x+1)^2}$

$$= \dfrac{3^x(x \ln 3 + \ln 3 - 1)}{(x+1)^2}$$

31. $\dfrac{dy}{dx} = -2x \cdot 2^{-x^2} \ln 2$

$$= 2^{-x^2}(-2x \ln 2)$$

33. $\dfrac{dA}{dt} = Pre^{rt}$

35a. $\dfrac{dy}{dx} = -100(-0.3e^{-0.3x})$

$= 30e^{-0.3x}$

$\dfrac{dy}{dx}\bigg|_{x=5} = 30e^{-1.5}$

≈ 6.69

35b. $\dfrac{dy}{dx}\bigg|_{x=10} = 30e^{-3}$

≈ 1.49

37a. $C'(x) = \dfrac{1}{2}e^x + \dfrac{3}{x+1}$

$C'(3) = \dfrac{1}{2}e^3 + \dfrac{3}{3+1}$

≈ 10.79

37b. $C'(10) = \dfrac{1}{2}e^{10} + \dfrac{3}{10+1}$

≈ 11013.5 (increasing)

37c. Yes, $C'(x)$ is always positive for $x \geq 0$.

39. $\dfrac{dn}{dx} = -50(-0.4)e^{-0.4x} = 20e^{-0.4x}$

$\dfrac{dn}{dx}\bigg|_{x=10} = 20e^{-4} \approx 0.37$

41. $N'(t) = (4-2t)(5000)e^{(4+4t-t^2)}$

$= 10000(2-t)e^{(4+4t-t^2)}$

Set $N'(t) = 0$:

$10000(2-t)e^{(4+4t-t^2)} = 0$

$2-t = 0$

$t = 2$

Using the first derivative test:

interval	$0 < t < 2$	$t > 2$
sign of $N'(t)$	$+$	$-$
location	left of 2	right of 2

maximum at $t = 2$ hours

43a. $N(0) = 100e^0 + 30$

$= 130$

43b. $N(1) = 100e^{-1} + 30$

≈ 67

43c. $N'(t) = -100e^{-t}$

$N'(6) = -100e^{-6}$

≈ -0.25

Section 5.5   (pages 360-362)

1. $f'(p) = -2p$

$$\frac{f'(10)}{f(10)} = \frac{-20}{300} = \frac{-1}{15}$$
$$\approx -0.0667$$

Demand is decreasing by about 6.67% of the total demand.

3. $f'(p) = -2p$

$$\frac{f'(15)}{f(15)} = \frac{-30}{175} = \frac{-6}{35}$$
$$\approx -0.1714$$

Demand is decreasing by about 17.14% of the total demand.

5a. $E(p) = \dfrac{pf'(p)}{f(p)}$

$$= \frac{p(-10)}{400 - 10p}$$

$$= \frac{p}{p - 40}$$

7a. $E(p) = \dfrac{pf'(p)}{f(p)}$

$$= \frac{p(-40)}{800 - 40p}$$

$$= \frac{p}{p - 20}$$

5b. $E(10) = \dfrac{10}{10 - 40} = \dfrac{-1}{3}$

7b. $E(10) = \dfrac{10}{10 - 20} = -1$

5c. $E(20) = \dfrac{20}{20 - 40} = -1$

7c. $E(5) = \dfrac{5}{5 - 20} = \dfrac{-1}{3}$

5d. $E(30) = \dfrac{30}{30 - 40} = -3$

7d. $E(15) = \dfrac{15}{15 - 20} = -3$

9a. $E(p) = \dfrac{pf'(p)}{f(p)}$

$$= \frac{p(-10p)}{4500 - 5p^2}$$

$$= \frac{2p^2}{p^2 - 900}$$

11a. $E = \dfrac{pf'(p)}{f(p)}$

$$= \frac{p(-4p)}{5000 - 2p^2}$$

$$= \frac{2p^2}{p^2 - 2500}$$

9b. $E(5) = \dfrac{2(5)^2}{(5)^2 - 900} = \dfrac{-2}{35}$

11b. $E(20) = \dfrac{2(20)^2}{(20)^2 - 2500} = \dfrac{-8}{21}$

9c. $E(20) = \dfrac{2(20)^2}{(20)^2 - 900} = \dfrac{-8}{5}$

11c. $E(40) = \dfrac{2(40)^2}{(40)^2 - 2500} = \dfrac{-32}{9}$

13a. $E(p) = \dfrac{pf'(p)}{f(p)}$

$$= \frac{p(-20p^{-3})}{10p^{-2}}$$

$$= -2$$

15a. $E(p) = \dfrac{pf'(p)}{f(p)}$

$$= \frac{p(-100e^{-p})}{100e^{-p}}$$

$$= -p$$

13b. $E(1) = -2$

15b. $E(2) = -2$

15c. $E\left(\dfrac{1}{2}\right) = \dfrac{-1}{2}$

17. $E(p) = \dfrac{p(-4)}{1200 - 4p}$

$ = \dfrac{p}{p - 300}$

$E(100) = \dfrac{100}{100 - 300} = \dfrac{-1}{2}$

inelastic

19. $E(p) = \dfrac{p}{p - 300}$

$E(200) = \dfrac{200}{200 - 300} = -2$

elastic

21. unit elasticity

23. inelastic

25a. dollars

25b. items

25c. items/dollar

25d. $\dfrac{(\text{dollars})\left(\dfrac{\text{items}}{\text{dollar}}\right)}{\text{items}}$ (All units cancel.)

27a. $R(p) = p(2700 - 3p)$

$ = 2700p - 3p^2$

$R'(p) = 2700 - 6p$

27b,c. Set $R'(p) = 0$:

$2700 - 6p = 0$

$p = 450$

interval	$0 < p < 450$	$450 < p < 900$
sign of $R'(p)$	$+$	$-$
result: $R(p)$ is	increasing	decreasing

27d,e. $E(p) = \dfrac{pf'(p)}{f(p)}$

$ = \dfrac{p(-3)}{2700 - 3p}$

$ = \dfrac{p}{p - 900}$

When demand is elastic:

$E(p) < -1$

$\dfrac{p}{p - 900} < -1$

$\dfrac{p}{p - 900} + 1 < 0$

$\dfrac{p + p - 900}{p - 900} < 0$

$\dfrac{2p - 900}{p - 900} < 0$

so $450 < p < 900$

When demand is inelastic:

$\dfrac{2p - 900}{p - 900} > 0$

so $0 < p < 450$

29a.
$$E(p) = \frac{pf'(p)}{f(p)}$$

$$= \frac{p(-b)}{a - bp}$$

$$= \frac{p}{p - \frac{a}{b}}$$

29b.

p	$E(p)$
$\frac{a}{4b}$	$\frac{-1}{3}$
$\frac{a}{3b}$	$\frac{-1}{2}$
$\frac{a}{2b}$	-1
$\frac{2a}{3b}$	-2
$\frac{10a}{11b}$	-10

31a. $a = 4800,\ b = 2,\ 0 < p < \frac{4800}{2}$

$p > \frac{4800}{2(2)}$

$p > 1200$

$1200 < p < 2400$

31b. $p < \frac{4800}{2(2)}$

$p < 1200$

$0 < p < 1200$

31c. $p = \frac{4800}{2(2)} = 1200$

33.
$$E(p) = \frac{p(-12p)}{5400 - 6p^2}$$

$$= \frac{2p^2}{p^2 - 900}$$

$$E(15) = \frac{2(15)^2}{(15)^2 - 900} = \frac{-2}{3}$$

$$\frac{-2}{3}(20\%) = -13\tfrac{1}{3}\%$$

Demand increase by $13\tfrac{1}{3}\%$.

Chapter 5 Review   (pages 362-364)

1.

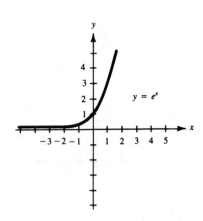

2.

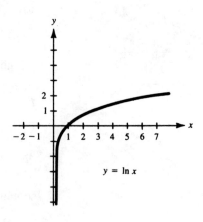

3. $\log_5 25 = 2$

4. $\log_{10} 0.01 = -2$

5. $\log_6 \frac{1}{36} = -2$

6. $\ln 1 = 0$

7. $10^2 = 100$

8. $e^{-1} = \frac{1}{e}$

9. $b^0 = 1$

10. $10^{-3} = 0.001$

11a. $\dfrac{dR}{dx} = 30x\left(\dfrac{-1}{4}\right)e^{-x/4} + 30e^{-x/4}$

$\qquad = 30e^{-x/4}\left(\dfrac{-1}{4}x + 1\right)$

Set $\dfrac{dR}{dx} = 0$: $\qquad 30e^{-x/4}\left(\dfrac{-1}{4}x + 1\right) = 0$

$\qquad\qquad\qquad\qquad \dfrac{-1}{4}x + 1 = 0$

$\qquad\qquad\qquad\qquad\qquad x = 4$

Using the second derivative test:

$\dfrac{d^2 R}{dx^2} = 30e^{-x/4}\left(\dfrac{-1}{4}\right) + 30\left(\dfrac{-1}{4}\right)e^{-x/4}\left(\dfrac{-1}{4}x + 1\right)$

$\qquad = \dfrac{-15}{2}e^{-x/4}\left(2 - \dfrac{1}{4}x\right)$

$\left.\dfrac{d^2 R}{dx^2}\right|_{x=4} \quad \dfrac{-15}{2}e^{-1}(2 - 1) \qquad$ maximum at $x = 4$

11b. $R = 30(4)e^{-4/4} \approx 44.15$

12. $R(x) = px$

$\qquad = 80xe^{-x/2}$

$R'(x) = 80x\left(\dfrac{-1}{2}\right)e^{-x/2} + 80e^{-x/2}$

$\qquad = -40e^{-x/2}(x - 2)$

Set $R'(x) = 0$: $\quad -40e^{-x/2}(x - 2) = 0$

$\qquad\qquad\qquad\qquad x - 2 = 0$

$\qquad\qquad\qquad\qquad\qquad x = 2$

Using the second derivative test:

$R''(x) = -40e^{-x/2}(1) + (-40)\left(-\dfrac{1}{2}\right)e^{-x/2}(x - 2)$

$\qquad = 20e^{-x/2}(x - 4)$

$R''(2) = 20e^{-1}(2 - 4) \qquad$ (maximum)

$R(2) = 80(2)e^{-1} \approx 58.86 \qquad$ Maximum revenue of \$58.86 at $x = 2$.

13a.

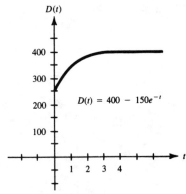

$D(t)$

400

300

200

$D(t) = 400 - 150e^{-t}$

100

1 2 3 4

t

13b. $D'(t) = 150e^{-t}$

$D'(2) = 150e^{-2} \approx 20.3$

$D'(4) = 150e^{-4} \approx 2.75$

$D'(10) = 150e^{-10} \approx 0.007$

14. $2N_0 = N_0(1.06)^t$

$2 = (1.06)^t$

$\log 2 = \log (1.06)^t$

$\log 2 = t \log (1.06)$

$t = \dfrac{\log 2}{\log (1.06)}$

$t \approx 11.9$ years

15. $\dfrac{dy}{dx} = -e^{-x}$

16. $\dfrac{dy}{dx} = \dfrac{1}{2x+1}(2) = \dfrac{2}{2x+1}$

17. $\dfrac{dy}{dx} = \dfrac{1}{4-x^2}(-2x) = \dfrac{-2x}{4-x^2}$

18. $\dfrac{dy}{dx} = -2xe^{-x^2}$

19. $\dfrac{dy}{dx} = x^3e^x + 3x^2e^x$

$= x^2e^x(x+3)$

20. $\dfrac{dy}{dx} = \dfrac{(x+1)\frac{1}{x} - \ln x(1)}{(x+1)^2}$

$= \dfrac{\frac{x+1}{x} + \frac{x \ln x}{x}}{(x+1)^2}$

$= \dfrac{x+1+x \ln x}{x(x+1)^2}$

21. $\dfrac{dy}{dx} = \dfrac{e^x - e^{-x}}{2}$

22. $y = \ln(x^2+1)^{1/2}$

$y = \frac{1}{2} \ln(x^2+1)$

$\dfrac{dy}{dx} = \frac{1}{2} \cdot \dfrac{1}{x^2+1}(2x)$

$= \dfrac{x}{x^2+1}$

23a. $R(p) = p(2700 - 6p)$

$= 2700p - 6p^2$

23b. $E(p) = \dfrac{pf'(p)}{f(p)}$

$= \dfrac{p(-6)}{2700 - 6p}$

$= \dfrac{p}{p - 450}$

23c,d. Demand is elastic when

$$E(p) < -1$$

$$\frac{p}{p - 450} < -1$$

$$\frac{p}{p - 450} + 1 < 0$$

$$\frac{2p - 450}{p - 450} < 0$$

so $225 < p < 450$

Demand is inelastic for $E(p) > -1$

so $0 < p < 225$

24a. $E(p) = \dfrac{pf'(p)}{f(p)}$

$$= \frac{p(-32p)}{4800 - 16p^2}$$

$$= \frac{2p^2}{p^2 - 300}$$

24b. $E(5) = \dfrac{2(5)^2}{(5)^2 - 300} = \dfrac{-2}{11}$

24c. $E(10) = \dfrac{2(10)^2}{(10)^2 - 300} = -1$

Chapter 5 Test (pages 364-366)

1. $\log_P R = Q$ (c)

2. d (since $\left(\frac{1}{2}\right)^{-x} = 2^x$ and $(0.1)^{-x} = 10^x$)

3. $\dfrac{dy}{dx} = 2 \ln x \left(\dfrac{1}{x}\right)$

$\qquad = \frac{2}{x} \ln x$ (c)

4. $(2^2)^{x-1} = \dfrac{1}{2^3}$

$\qquad 2^{2x-2} = 2^{-3}$

$\qquad 2x - 2 = -3$

$\qquad 2x = -1$

$\qquad x = \dfrac{-1}{2}$ (a)

5. $4^1 = \sqrt{x}$, $16 = x$ (b)

6. $\dfrac{dy}{dx} = \dfrac{(e^{2x} + 1)(2e^{2x}) - e^{2x}(2e^{2x})}{(e^{2x} + 1)^2}$

$\qquad = \dfrac{2e^{4x} + 2e^{2x} - 2e^{4x}}{(e^{2x} + 1)^2}$

$\qquad = \dfrac{2e^{2x}}{(e^{2x} + 1)^2}$ (b)

7. $\log_x 9 = -2$

$\qquad x^{-2} = 9$

$\qquad \dfrac{1}{x^2} = 9$

$\qquad x^2 = \dfrac{1}{9}$

$\qquad x = \dfrac{1}{3}$ (d)

8. a (vertical asymptote $x = 1$, increasing)

9. $\ln e^{2x+1} = \ln 7$

$(2x+1)\ln e = \ln 7$

$2x + 1 = \ln 7$

$2x = -1 + \ln 7$

$x = \frac{-1}{2} + \frac{1}{2}\ln 7$ (c)

10. $\frac{dy}{dx} = x^2\left(\left(\frac{1}{x}\right)\right) + 2x \ln x$

$= x(1 + 2\ln x)$

$1 + 2\ln x = 0$

$\ln x = \frac{-1}{2}$

$e^{-1/2} = x$

Using the second derivative test:

$\frac{d^2y}{dx^2} = x\left(\frac{2}{x}\right) + (1)(1 + 2\ln x)$

$= 3 + 2\ln x$

$\left.\frac{d^2y}{dx^2}\right|_{x = e^{-1/2}} = 3 + 2\ln e^{-1/2} = 2$

(minimum)

minimum point $\left(\frac{1}{\sqrt{e}}, \frac{-1}{2e}\right)$ (b)

11. $e^{1-2x} = \frac{12}{5}$

$\ln e^{1-2x} = \ln 2.4$

$(1 - 2x)\ln e = \ln 2.4$

$1 - 2x = \ln 2.4$

$-2x = -1 + \ln 2.4$

$x = \frac{-1 + \ln 2.4}{-2}$

$x \approx 0.0623$

12.

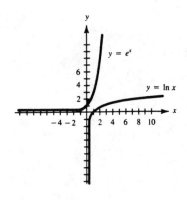

$y = e^x$

6

4

$y = \ln x$

2

$-4 \ -2 \quad 2 \ 4 \ 6 \ 8 \ 10$

13. $\frac{dy}{dx} = 2xe^{x^2-1} + 10 = 2(xe^{x^2-1} + 5)$

14. $f'(x) = \frac{1}{x^2+1}(2x)$

$= \frac{2x}{x^2+1}$

15. $\frac{dy}{dx} = \frac{x\left(\frac{1}{x}\right) - \ln x(1)}{x^2}$

$= \frac{1 - \ln x}{x^2}$

16. $f'(x) = xe^x + (1)e^x$

$= e^x(x+1)$

Set $f'(x) = 0$: $e^x(x+1) = 0$

$x + 1 = 0$

$x = -1$

17. $f'(x) = 2(-2x)e^{-x^2}$

$\qquad = -4xe^{-x^2}$

Set $f'(x) = 0$:

$-4xe^{-x^2} = 0$

$-4x = 0$

$x = 0$

Using the first derivative test:

interval	$x < 0$	$x > 0$
sign of $f'(x)$	$+$	$-$
location	left of 0	right of 0

absolute maximum: $f(0) = 2e^0 = 2$

18. $V = 45000(1.08)^5 \approx \$66,119.76$

19a. $\dfrac{dy}{dx} = -150(-0.2)e^{-0.2x}$

$\qquad = 30e^{-0.2x}$

$\dfrac{dy}{dx}\bigg|_{x=5} = 30e^{-0.2(5)}$

$\qquad = 30e^{-1}$

$\qquad \approx 11.04$

19b. $\dfrac{dy}{dx}\bigg|_{x=10} = 30e^{-0.2(10)}$

$\qquad = 30e^{-2}$

$\qquad \approx 4.06$

20a. $R(x) = px$

$\qquad = (1000e^{-0.01x})x$

$\qquad = 1000xe^{-0.01x}$

20b. $R'(x) = 1000x(-0.01)e^{-0.01x} + 1000e^{-0.01x}$

$\qquad = -10xe^{-0.01x} + 1000e^{-0.01x}$

$\qquad = -10e^{-0.01x}(x - 100)$

CHAPTER 6: INTEGRATION

Section 6.1 (pages 375-379)

1. $5x + C$

3. $4\displaystyle\int x^3\,dx = 4 \cdot \dfrac{x^4}{4} + C$

$\qquad = x^4 + C$

5. $4\displaystyle\int x^5\,dx = 4\dfrac{x^6}{6} + C$

$\qquad = \tfrac{2}{3}x^6 + C$

7. $\displaystyle\int (3x^2 - 2x + 5)\,dx = 3 \cdot \dfrac{x^3}{3} - 2 \cdot \dfrac{x^2}{2} + 5x + C$

$\qquad = x^3 - x^2 + 5x + C$

9. $\int (6y^5 - 30y^2 + 7y - 4)dy = 6 \cdot \frac{y^6}{6} - 30 \cdot \frac{y^3}{3} + 7 \cdot \frac{y^2}{2} - 4y + C$
$$= y^6 - 10y^3 + \frac{7}{2}y^2 - 4y + C$$

11. $\int (3x^6 - 4x + 7)dx = 3 \cdot \frac{x^7}{7} - 4 \cdot \frac{x^2}{2} + 7x + C$
$$= \frac{3}{7}x^7 - 2x^2 + 7x + C$$

13. $\int \frac{1}{x}dx = \ln|x| + C$

15. $\int \left[5 - 3\left(\frac{1}{x}\right)\right]dx = 5x - 3\ln|x| + C$

17. $\frac{1}{4}\int (20x^3 - 6x + 1)dx = \frac{1}{4}\left(20 \cdot \frac{x^4}{4} - 6 \cdot \frac{x^2}{2} + x\right) + C$
$$= \frac{5}{4}x^4 - \frac{3}{4}x^2 + \frac{1}{4}x + C$$

19. $\int (30x^{-3} - 2x^{-2} - \frac{1}{x})dx = 30 \cdot \frac{x^{-2}}{-2} - 2 \cdot \frac{x^{-1}}{-1} - \ln|x| + C$
$$= \frac{-15}{x^2} + \frac{2}{x} - \ln|x| + C$$

21. $\int (x - e^x)dx = \frac{x^2}{2} - e^x + C$

23. $\int (x^{1/3} - 4x^{1/2})dx = \frac{x^{4/3}}{\frac{4}{3}} - 4 \cdot \frac{x^{3/2}}{\frac{3}{2}} + C = \frac{3\sqrt[3]{x^4}}{4} - \frac{8\sqrt{x^3}}{3} + C$

25. $\int 20x^{-3/4}dx = 20 \cdot \frac{x^{1/4}}{\frac{1}{4}} + C = 80\sqrt[4]{x} + C$

27. $\int \frac{x}{x^{2/3}}dx = \int x^{1/3}dx = \frac{x^{4/3}}{\frac{4}{3}} + C = \frac{3\sqrt[3]{x^4}}{4} + C$

29. $\int (4x^2 + 4x + 1)dx = 4 \cdot \frac{x^3}{3} + 4 \cdot \frac{x^2}{2} + x + C = \frac{4}{3}x^3 + 2x^2 + x + C$

31. $f(x) = \int f'(x)dx$

$f(x) = \int (3x^2 - 6x + 1)dx$

$f(x) = 3 \cdot \frac{x^3}{3} - 6 \cdot \frac{x^2}{2} + x + C$

$f(x) = x^3 - 3x^2 + x + C$

$f(0) = (0)^3 - 3(0)^2 + 0 + C$

$-10 = C$

$f(x) = x^3 - 3x^2 + x - 10$

33. $f(x) = \int f'(x)dx$

$f(x) = \int (8x^3 - x^2 + x)dx$

$f(x) = 8 \cdot \frac{x^4}{4} - \frac{x^3}{3} + \frac{x^2}{2} + C$

$f(x) = 2x^4 - \frac{x^3}{3} + \frac{x^2}{2} + C$

$f(0) = 2(0)^2 - \frac{(0)^3}{3} + \frac{(0)^2}{2} + C$

$10 = C$

$f(x) = 2x^4 - \frac{x^3}{3} + \frac{x^2}{2} + 10$

35.　$f(x) = \displaystyle\int f'(x)dx$

　　$f(x) = \displaystyle\int \left[3\!\left(\tfrac{1}{x}\right) - 5\right]dx$

　　$f(x) = 3\ln|x| - 5x + C$

　　$f(1) = 3\ln|1| - 5(1) + C$

　　$0 = 0 - 5 + C$

　　$5 = C$

　　$f(x) = 3\ln|x| - 5x + 5$

37a.　$f(x) = \tfrac{1}{2}x^2 + 2$　　　　$f(x) = \tfrac{1}{2}x^2$　　　　$f(x) = \tfrac{1}{2}x^2 - 4$

　　　$f'(x) = x$　　　　　　　　$f'(x) = x$　　　　　　$f'(x) = x$

　　　$f'(2) = 2$　　　　　　　　$f'(2) = 2$　　　　　　$f'(2) = 2$

　　　The slope of the tangent line is 2.

37b.

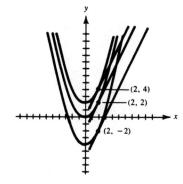

(2, 4)
(2, 2)
(2, −2)

39a.　$C(x) = \displaystyle\int C'(x)dx$

　　　$C(x) = \displaystyle\int (2x - 6)dx$

　　　$C(x) = 2 \cdot \dfrac{x^2}{2} - 6x + k$

　　　$C(x) = x^2 - 6x + k$

　　　$C(0) = (0)^2 - 6(0) + k$

　　　$400 = k$

　　　$C(x) = x^2 - 6x + 400$

39b.　$C(50) = (50)^2 - 6(50) + 400 = \$2600$

41. $C(x) = \displaystyle\int C'(x)dx$

   $C(x) = \displaystyle\int (600x^{-1/3} + 0.1x)dx$

   $C(x) = 600 \cdot \dfrac{x^{2/3}}{\frac{2}{3}} + (0.1)\dfrac{x^2}{2} + k$

   $C(x) = 900\sqrt[3]{x^2} + 0.05x^2 + k$

   $C(0) = 900\sqrt[3]{0^2} + 0.05(0)^2 + k$

   $20,000 = k$

   $C(x) = 900\sqrt[3]{x^2} + 0.05x^2 + 20,000$

   $C(1000) = 900\left(\sqrt[3]{1000}\right)^2 + 0.05(1000)^2 + 20,000$

   $ = \$160,000$

43. $S(t) = \displaystyle\int S'(t)dt$

   $S(t) = \displaystyle\int 0.8t\,dt$

   $S(t) = (0.8)\dfrac{t^2}{2} + C$

   $S(t) = 0.4t^2 + C$

   $S(5) = 0.4(5)^2 + C$

   $10 = 10 + C$

   $0 = C$

   $S(t) = 0.4t^2$

45. $P(x) = \displaystyle\int P'(x)dx$

   $P(x) = \displaystyle\int (3x^2 - 360x + 10,800)dx$

   $P(x) = 3 \cdot \dfrac{x^3}{3} - 360\dfrac{x^2}{2} + 10,800x + C$

   $P(x) = x^3 - 180x^2 + 10,800x + C$

   $P(50) = (50)^3 - 180(50)^2 + 10,800(50) + C$

   $-200 = 215,000 + C$

   $-215,200 = C$

   $P(x) = x^3 - 180x^2 + 10,800x - 215,200$

47.

$$N(t) = \int N'(t)\,dt$$

$$N(t) = \int (-1000t)\,dt$$

$$N(t) = -1000\frac{t^2}{2} + C$$

$$N(t) = -500t^2 + C$$

$$N(0) = -500(0)^2 + C$$

$$80,000 = C$$

$$N(t) = -500t^2 + 80,000$$

47b.

$$0 = -500t^2 + 80,000$$

$$500t^2 = 80,000$$

$$t^2 = 160$$

$$t \approx 12.6 \text{ months}$$

49.

$$P(t) = \int P'(t)\,dt$$

$$P(t) = \int (2 + 5t^{1/3})\,dt$$

$$P(t) = 2t + 5\frac{t^{4/3}}{\frac{4}{3}} + C$$

$$P(t) = 2t + \frac{15}{4}t^{4/3} + C$$

49b.

$$P(0) = 2(0) + \frac{15}{4}(0)^{4/3} + C$$

$$2600 = C$$

$$P(t) = 2t + \frac{15}{4}t^{4/3} + 2600$$

$$P(60) = 2(60) + \frac{15}{4}\left(\sqrt[3]{60}\right)^4 + 2600$$

$$\approx 3601$$

51.

$$v(t) = \int a(t)\,dt$$

$$v(t) = \int (2t^2 + 2t)\,dt$$

$$v(t) = 2 \cdot \frac{t^3}{3} + 2\frac{t^2}{2} + C_1$$

$$v(t) = \frac{2}{3}t^3 + t^2 + C_1$$

$$v(0) = \frac{2}{3}(0)^3 + (0)^2 + C_1$$

$$5 = C_1$$

$$v(t) = \frac{2}{3}t^3 + t^2 + 5$$

$$s(t) = \int v(t)\,dt$$

$$s(t) = \int \left(\frac{2}{3}t^3 + t^2 + 5\right)dt$$

$$s(t) = \frac{2}{3} \cdot \frac{t^4}{4} + \frac{t^3}{3} + 5t + C_2$$

$$s(t) = \frac{1}{6}t^4 + \frac{1}{3}t^3 + 5t + C_2$$

$$s(0) = \frac{1}{6}(0)^4 + \frac{1}{3}(0)^3 + 5(0) + C_2$$

$$0 = C_2$$

$$s(t) = \frac{1}{6}t^4 + \frac{1}{3}t^3 + 5t$$

53a.

$$v(t) = \int a(t)\,dt$$

$$v(t) = \int (-32)\,dt$$

$$v(t) = -32t + C_1$$

$$v(0) = -32(0) + C_1$$

$$0 = C_1$$

$$v(t) = -32t$$

$$s(t) = \int v(t)\,dt$$

$$s(t) = \int (-32t)\,dt$$

$$s(t) = -32 \cdot \frac{t^2}{2} + C_2$$

$$s(t) = -16t^2 + C_2$$

$$s(0) = -16(0)^2 + C_2$$

$$160 = C_2$$

$$s(t) = -16t^2 + 160$$

53b. $0 = -16t^2 + 160$

$16t^2 = 160$

$t^2 = 10$

$t = \sqrt{10} \approx 3.16$ sec.

53c. $v(\sqrt{10}) = -32(\sqrt{10})$

≈ -101.2 ft./sec.

Section 6.2 (pages 386-388)

1. $\int (x+3)^5 dx$ $\boxed{\begin{aligned} u &= x+3 \\ du &= dx \end{aligned}}$

$= \int u^5 du$

$= \dfrac{u^6}{6} + C$

$= \dfrac{1}{6}(x+3)^6 + C$

3. $\int (4x-7)^3 dx$ $\boxed{\begin{aligned} u &= 4x-7 \\ du &= 4dx \end{aligned}}$

$= \dfrac{1}{4}\int (4x-7)^3(4dx)$

$= \dfrac{1}{4}\int u^3 du$

$= \dfrac{1}{4}\cdot\dfrac{u^4}{4} + C$

$= \dfrac{1}{16}(4x-7)^4 + C$

5. $\int (x^3+1)^5(3x^2 dx)$ $\boxed{\begin{aligned} u &= x^3+1 \\ du &= 3x^2 dx \end{aligned}}$

$= \int u^5 du$

$= \dfrac{u^6}{6} + C$

$= \dfrac{1}{6}(x^3+1)^6 + C$

7. $\int (x^5-1)^{1/2}(5x^4 dx)$ $\boxed{\begin{aligned} u &= x^5-1 \\ du &= 5x^4 dx \end{aligned}}$

$= \int u^{1/2} du$

$= \dfrac{u^{3/2}}{\frac{3}{2}} + C$

$= \dfrac{2}{3}\sqrt{(x^5-1)^3} + C$

9. $\int (y^3-1)^{1/3}(3y^2 dy)$ $\boxed{\begin{aligned} u &= y^3-1 \\ du &= 3y^2 dy \end{aligned}}$

$= \int u^{1/3} du$

$= \dfrac{u^{4/3}}{\frac{4}{3}} + C$

$= \dfrac{3}{4}(y^3-1)^{4/3} + C$

11. $\displaystyle\int (x^3-1)^{-1/2}x^2dx$

$\qquad\boxed{\begin{array}{l} u=x^3-1 \\[4pt] du=3x^2dx \end{array}}$

$\displaystyle =\tfrac{1}{3}\int (x^3-1)^{-1/2}(3x^2dx)$

$\displaystyle =\tfrac{1}{3}\int u^{-1/2}du$

$\displaystyle =\tfrac{1}{3}\cdot\frac{u^{1/2}}{\frac{1}{2}}+C$

$\displaystyle =\tfrac{2}{3}\sqrt{x^3-1}+C$

13. $\displaystyle\int \frac{1}{x^3-1}(x^2dx)$

$\qquad\boxed{\begin{array}{l} u=x^3-1 \\[4pt] du=3x^2dx \end{array}}$

$\displaystyle =\tfrac{1}{3}\int \frac{1}{x^3-1}(3x^2dx)$

$\displaystyle =\tfrac{1}{3}\int \tfrac{1}{u}du$

$\displaystyle =\tfrac{1}{3}\ln|u|+C$

$\displaystyle =\tfrac{1}{3}\ln|x^3-1|+C$

15. $\displaystyle\int e^{x^3}x^2dx$

$\qquad\boxed{\begin{array}{l} u=x^3 \\[4pt] du=3x^2dx \end{array}}$

$\displaystyle =\tfrac{1}{3}\int e^{x^3}(3x^2dx)$

$\displaystyle =\tfrac{1}{3}\int e^udu$

$\displaystyle =\tfrac{1}{3}e^u+C$

$\displaystyle =\tfrac{1}{3}e^{x^3}+C$

17. $\displaystyle\int \frac{1}{3t^2-1}(tdt)$

$\qquad\boxed{\begin{array}{l} u=3t^2-1 \\[4pt] du=6tdt \end{array}}$

$\displaystyle =\tfrac{1}{6}\int \frac{1}{3t^2-1}(6tdt)$

$\displaystyle =\tfrac{1}{6}\int \tfrac{1}{u}du$

$\displaystyle =\tfrac{1}{6}\ln|u|+C$

$\displaystyle =\tfrac{1}{6}\ln|3t^2-1|+C$

19. $\displaystyle\int \frac{1}{x^2-4}(x\,dx)$

$$u = x^2 - 4$$
$$du = 2x\,dx$$

$$=\frac{1}{2}\int \frac{1}{x^2-4}(2x\,dx)$$

$$=\frac{1}{2}\int \frac{1}{u}\,du$$

$$=\frac{1}{2}\ln|u| + C$$

$$=\frac{1}{2}\ln|x^2-4| + C$$

21. $\displaystyle\int e^{-x^2}x\,dx$

$$u = -x^2$$
$$du = -2x\,dx$$

$$=\frac{-1}{2}\int e^{-x^2}(-2x\,dx)$$

$$=\frac{-1}{2}\int e^{u}\,du$$

$$=\frac{-1}{2}e^{u} + C$$

$$=\frac{-1}{2}e^{-x^2} + C$$

23. $\displaystyle 5\int (2x-1)^{1/2}\,dx$

$$u = 2x - 1$$
$$du = 2\,dx$$

$$=\frac{5}{2}\int (2x-1)^{1/2}(2\,dx)$$

$$=\frac{5}{2}\int u^{1/2}\,du$$

$$=\frac{5}{2}\cdot\frac{u^{3/2}}{\frac{3}{2}} + C$$

$$=\frac{5}{3}\sqrt{(2x-1)^3} + C$$

25. $\displaystyle\int \frac{1}{x^3-3x^2+5}(3x^2-6x)\,dx$

$$u = x^3 - 3x^2 + 5$$
$$du = (3x^2 - 6x)\,dx$$

$$=\int \frac{1}{u}\,du$$

$$= \ln|u| + C$$

$$= \ln|x^3 - 3x^2 + 5| + C$$

27. $\int \frac{1}{x^3-5x^2+7x-1}(x^2-\frac{10}{3}x+\frac{7}{3})dx$

$$\boxed{\begin{array}{l} u = x^3-5x^2+7x-1 \\ du = (3x^2-10x+7)dx \end{array}}$$

$=\frac{1}{3}\int \frac{1}{x^3-5x^2+7x-1}(3x^2-10x+7)dx$

$=\frac{1}{3}\int \frac{1}{u}du$

$=\frac{1}{3}\cdot \ln|u|+C$

$=\frac{1}{3}\ln|x^3-5x^2+7x-1|+C$

29. $\int e^{-0.2x^2}xdx$

$$\boxed{\begin{array}{l} u = -0.2x^2 \\ du = -0.4xdx \\ \quad = \frac{-2}{5}xdx \end{array}}$$

$=\frac{-5}{2}\int e^{-0.2x^2}(\frac{-2}{5}xdx)$

$=\frac{-5}{2}\int e^u du$

$=\frac{-5}{2}e^u+C$

$=-2.5e^{-0.2x^2}+C$

31. $\int (1+0.5x)^4 dx$

$$\boxed{\begin{array}{l} u = 1+0.5x \\ du = 0.5dx \end{array}}$$

$=2\int (1+0.5x)^4(0.5dx)$

$=2\int u^4 du$

$=2\frac{u^5}{5}+C$

$=0.4(1+0.5x)^5+C$

33. $\int e^{x^{1/2}}x^{-1/2}dx$

$$\boxed{\begin{array}{l} u = x^{1/2} \\ du = \frac{1}{2}x^{-1/2}dx \end{array}}$$

$=2\int e^{x^{1/2}}(\frac{1}{2}x^{-1/2}dx)$

$=2\int e^u du$

$=2e^u+C$

$=2e^{\sqrt{x}}+C$

35.

$$\int (x^3 - 5x)^{2/3}(3x^2 - 5)dx$$

$$\boxed{\begin{aligned} u &= x^3 - 5x \\ du &= (3x^2 - 5)dx \end{aligned}}$$

$$= \int u^{2/3}du$$

$$= \frac{u^{5/3}}{\frac{5}{3}} + C$$

$$= \frac{3}{5}(x^3 - 5x)^{5/3} + C$$

37.

$$\int \frac{1}{x^2 + 7x + 12}(6x + 21)dx$$

$$\boxed{\begin{aligned} u &= x^2 + 7x + 12 \\ du &= (2x + 7)dx \end{aligned}}$$

$$= 3\int \frac{1}{x^2 + 7x + 12}(2x + 7)dx$$

$$= 3\int \frac{1}{u}du$$

$$= 3\ln|u| + C$$

$$= 3\ln|x^2 + 7x + 12| + C$$

39.

$$\int (1 - x^{1/2})^{-3}x^{-1/2}dx$$

$$\boxed{\begin{aligned} u &= 1 - x^{1/2} \\ du &= \frac{-1}{2}x^{-1/2}dx \end{aligned}}$$

$$= -2\int (1 - x^{1/2})^{-3}\left(\frac{-1}{2}x^{-1/2}dx\right)$$

$$= -2\int u^{-3}du$$

$$= -2\frac{u^{-2}}{-2} + C$$

$$= \frac{1}{(1 - \sqrt{x})^2} + C$$

41.

$$\int (\ln x)\left(\frac{1}{x}dx\right)$$

$$\boxed{\begin{aligned} u &= \ln x \\ du &= \frac{1}{x}dx \end{aligned}}$$

$$= \int u\,du$$

$$= \frac{u^2}{2} + C$$

$$= \frac{(\ln x)^2}{2} + C$$

43.

$$\int (\ln 2x)\left(\frac{1}{x}dx\right)$$

$$\boxed{\begin{aligned} u &= \ln 2x \\ du &= \frac{1}{2x} \cdot 2dx \\ &= \frac{1}{x}dx \end{aligned}}$$

$$= \int u\,du$$

$$= \frac{u^2}{2} + C$$

$$= \frac{(\ln 2x)^2}{2} + C$$

45. $\frac{dy}{dx} = \frac{1}{8}(8)(2x^2-1)^7(4x) + 0 = 4x(2x^2-1)^7$

47. $\frac{dy}{dx} = \frac{2}{9}\left(\frac{3}{2}\right)(x^3+1)^{1/2}(3x^2) + 0 = x^2(x^3+1)^{1/2}$

49. $\frac{dy}{dx} = \frac{-1}{2}\cdot\frac{1}{1-x^2}(-2x) + 0 = \frac{x}{1-x^2}$

51a. $\displaystyle\int x(x-2)^{1/2}dx$

$\begin{array}{|c|}\hline u = x-2 \\ du = dx \\ u+2 = x \\ \hline\end{array}$

$= \displaystyle\int (u+2)u^{1/2}du$

$= \displaystyle\int (u^{3/2} + 2u^{1/2})du$

$= \dfrac{u^{5/2}}{\frac{5}{2}} + 2\cdot\dfrac{u^{3/2}}{\frac{3}{2}} + C$

$= \frac{2}{5}(x-2)^{5/2} + \frac{4}{3}(x-2)^{3/2} + C$

51b. $\displaystyle\int (x+2)^{-1/2}dx$

$\begin{array}{|c|}\hline u = x+2 \\ du = dx \\ u-2 = x \\ \hline\end{array}$

$= \displaystyle\int u^{-1/2}(u-2)du$

$= \displaystyle\int (u^{1/2} - 2u^{-1/2})du$

$= \dfrac{u^{3/2}}{\frac{3}{2}} - 2\cdot\dfrac{u^{1/2}}{\frac{1}{2}} + C$

$= \frac{2}{3}(x+2)^{3/2} - 4(x+2)^{1/2} + C$

51c. $\displaystyle\int (x+2)^{-1}x\,dx$

$\begin{array}{|c|}\hline u = x+2 \\ du = dx \\ u-2 = x \\ \hline\end{array}$

$= \displaystyle\int u^{-1}(u-2)du$

$= \displaystyle\int (1 - 2u^{-1})du$

$= u - 2\ln|u| + C_1$

$= x + 2 - 2\ln|x+2| + C_1$

$= x - 2\ln|x+2| + C$

51d. $\int (x-2)^{-4} x\, dx$

$\boxed{\begin{aligned} u &= x-2 \\ du &= dx \\ u+2 &= x \end{aligned}}$

$= \int u^{-4}(u+2)du$

$= \int (u^{-3} + 2u^{-4})du$

$= \dfrac{u^{-2}}{-2} + 2 \cdot \dfrac{u^{-3}}{-3} + C$

$= \dfrac{-1}{2(x-2)^2} - \dfrac{2}{3(x-2)^3} + C$

53. $C(x) = \int C'(x)dx$

$\boxed{\begin{aligned} u &= x^2 + 90{,}000 \\ du &= 2x\, dx \end{aligned}}$

$C(x) = \int (x^2 + 90{,}000)^{-1/2} x\, dx$

$C(x) = \frac{1}{2}\int (x^2 + 90{,}000)^{-1/2}(2x\, dx)$

$C(u) = \frac{1}{2}\int u^{-1/2} du$

$= \dfrac{1}{2}\dfrac{u^{1/2}}{\frac{1}{2}} + k$

$C(x) = \sqrt{x^2 + 90{,}000} + k$

$C(0) = \sqrt{0 + 90{,}000} + k$

$3500 = 300 + k$

$3200 = k$

$C(x) = \sqrt{x^2 + 90{,}000} + 3200$

$C(400) = \sqrt{(400)^2 + 90{,}000} + 3200 = \$3700$

55. $V(t) = \int V'(t)dt$

$= \int -560te^{-0.2t^2} dt$

$\boxed{\begin{aligned} u &= -0.2t^2 \\ du &= -0.4t\, dt \end{aligned}}$

$= -560 \int e^{-0.2t^2} t\, dt$

$= \dfrac{-5}{2}(-560)\int e^{-0.2t^2}(-0.4t\, dt)$

$V(u) = 1400 \int e^u du = 1400e^u + C$

$V(t) = 1400e^{-0.2t^2} + C$

$V(0) = 1400e^0 + C$

$4200 = 1400 + C,\ 2800 = C$

$V(t) = 1400e^{-0.2t^2} + 2800$

$V(3) = 1400e^{-0.2(9)} + 2800 \approx \$3031.42$

57.

$$R(t) = \int R'(t)dt$$

$$= \int -500e^{-0.4t}dt$$

$$= -500 \int e^{-0.4t}dt \qquad \boxed{u = -0.4t}$$

$$= \frac{-5}{2}(-500) \int e^{-0.4t}(-0.4dt) \qquad \boxed{du = -0.4dt}$$

$$R(u) = 1250 \int e^u du = 1250e^u + C$$

$$R(t) = 1250e^{-0.4t} + C$$

$$R(0) = 1250e^0 + C$$

$$2000 = 1250 + C, \ 750 = C$$

$$R(t) = 1250e^{-0.4t} + 750$$

$$R(10) = 1250e^{-4} + 750 \approx 772.9 \text{ g.}$$

Section 6.3 (pages 397-401)

1.
$$\int_1^4 (2x+3)dx$$
$$= x^2 + 3x \Big]_1^4$$
$$= [4^2 + 3(4)] - [1^2 + 3(1)]$$
$$= 28 - 4 = 24$$

3.
$$\int_0^4 3x^2 dx$$
$$= x^3 \Big]_0^4$$
$$= 4^3 - 0^3 = 64$$

5.
$$\int_0^2 3x^2 dx$$
$$= x^3 \Big]_0^2$$
$$= 2^3 - 0^3 = 8$$

7.
$$\int_1^2 (6x^2 - 2x + 1)dx$$
$$= 2x^3 - x^2 + x \Big]_1^2$$
$$= [2(2)^3 - (2)^2 + 2] - [2(1)^3 - (1)^2 + 1]$$
$$= 14 - 2 = 12$$

9.
$$\int_1^2 (3x^2 - 2x + 3)dx$$
$$= x^3 - x^2 + 3x \Big]_1^2$$
$$= [(2)^3 - (2)^2 + 3(2)] - [(1)^3 - (1)^2 + 3(1)]$$
$$= 10 - 3 = 7$$

11. $\displaystyle\int_0^1 e^{x^2}x\,dx$

$\qquad\qquad u = x^2$

$=\dfrac{1}{2}\displaystyle\int_0^1 e^{x^2}(2x\,dx)$

$\qquad\qquad du = 2x\,dx$

$=\dfrac{1}{2}\displaystyle\int_0^1 e^u\,du$

$\qquad\qquad$ if $x = 1$, $u = 1$

$\qquad\qquad$ if $x = 0$, $u = 0$

$=\dfrac{1}{2}e^u\Big]_0^1$

$=\dfrac{1}{2}\big[e^1 - e^0\big]$

$=\dfrac{e-1}{2}$

13. $\displaystyle\int_1^3 \dfrac{1}{x}dx$

$= \ln|x|\ \Big]_1^3$

$= \ln 3 - \ln 1$

$= \ln 3$

15. $\displaystyle\int_0^2 x^3\,dx$

$=\dfrac{x^4}{4}\Big]_0^2$

$=\dfrac{(2)^4}{4} - \dfrac{(0)^4}{4}$

$=4$

17. $\displaystyle\int_{-2}^{-1} x^{-2}\,dx$

$=\dfrac{x^{-1}}{-1}\bigg]_{-2}^{-1}$

$=\dfrac{-1}{x}\bigg]_{-2}^{-1}$

$=\left(\dfrac{-1}{-1}\right) - \left(\dfrac{-1}{-2}\right)$

$=1 - \dfrac{1}{2} = \dfrac{1}{2}$

19. $\displaystyle\int_1^4 x^{-1/2}\,dx$

$=\dfrac{x^{1/2}}{\frac{1}{2}}\bigg]_1^4$

$=2\sqrt{x}\ \big]_1^4$

$=2\sqrt{4} - 2\sqrt{1}$

$=4 - 2 = 2$

21. $\displaystyle\int_0^1 (2x+1)^{-2}\,dx$

$\qquad\qquad u = 2x + 1$

$=\dfrac{1}{2}\displaystyle\int_0^1 (2x+1)^{-2}(2\,dx)$

$\qquad\qquad du = 2\,dx$

$=\dfrac{1}{2}\displaystyle\int_1^3 u^{-2}\,du$

$\qquad\qquad$ if $x = 1$, $u = 3$

$\qquad\qquad$ if $x = 0$, $u = 1$

$=\dfrac{1}{2}\dfrac{u^{-1}}{-1}\bigg]_1^3$

$=\dfrac{-1}{2u}\bigg]_1^3$

$=\left(\dfrac{-1}{2(3)}\right) - \left(\dfrac{-1}{2(1)}\right) = \dfrac{1}{3}$

23. $\displaystyle\int_2^4 \frac{1}{x^2-1}x\,dx$

$\boxed{\begin{array}{l} u = x^2 - 1 \\[4pt] du = 2x\,dx \\[4pt] \text{if } x = 4,\; u = 15 \\[4pt] \text{if } x = 2,\; u = 3 \end{array}}$

$=\dfrac{1}{2}\displaystyle\int_2^4 \frac{1}{x^2-1}(2x\,dx)$

$=\dfrac{1}{2}\displaystyle\int_3^{15} \frac{1}{u}\,du$

$=\dfrac{1}{2}\ln|u|\;\Big]_3^{15}$

$=\dfrac{1}{2}(\ln 15 - \ln 3)$

$=\dfrac{1}{2}\ln \dfrac{15}{3}$

$=\dfrac{1}{2}\ln 5$

25. $\displaystyle\int_{-2}^0 (1-4x)^{1/2}\,dx$

$\boxed{\begin{array}{l} u = 1 - 4x \\[4pt] du = -4\,dx \\[4pt] \text{if } x = 0,\; u = 1 \\[4pt] \text{if } x = -2,\; u = 9 \end{array}}$

$=\dfrac{-1}{4}\displaystyle\int_{-2}^0 (1-4x)^{1/2}(-4\,dx)$

$=\dfrac{-1}{4}\displaystyle\int_9^1 u^{1/2}\,du$

$=\dfrac{-1}{4}\dfrac{u^{3/2}}{\frac{3}{2}}\Big]_9^1$

$=\dfrac{-1}{6}\left(\sqrt{u}\right)^3\Big]_9^1$

$=\dfrac{-1}{6}\left[(\sqrt{1})^3 - (\sqrt{9})^3\right]$

$=\dfrac{-1}{6}(1-27) = \dfrac{13}{3}$

27. $\displaystyle\int_0^2 (x^2-1)^4 x\,dx$

$\boxed{\begin{array}{l} u = x^2 - 1 \\[4pt] du = 2x\,dx \\[4pt] \text{if } x = 2,\; u = 3 \\[4pt] \text{if } x = 0,\; u = -1 \end{array}}$

$=\dfrac{1}{2}\displaystyle\int_0^2 (x^2-1)^4(2x\,dx)$

$=\dfrac{1}{2}\displaystyle\int_{-1}^3 u^4\,du$

$=\dfrac{1}{2}\dfrac{u^5}{5}\Big]_{-1}^3$

$=\dfrac{1}{10}\left[(3)^5 - (-1)^5\right]$

$=\dfrac{244}{10} = \dfrac{122}{5}$

29. $\displaystyle\int_{-1}^{1}(1-2x)^3 dx$

$$=\frac{-1}{2}\int_{-1}^{1}(1-2x)^3(-2dx)$$

$$=\frac{-1}{2}\int_{3}^{-1}u^3 du$$

$$=\frac{-1}{2}\frac{u^4}{4}\Big]_3^{-1}$$

$$=\frac{-1}{8}\big[(-1)^4-(3)^4\big]$$

$$=\frac{-1}{8}(-80)=10$$

$u=1-2x$
$du=-2dx$
if $x=1,\ u=-1$
if $x=-1,\ u=3$

31. $\displaystyle\int_{0}^{1}(e^x+1)e^x dx$

$$=\int_{2}^{e+1}u\,du$$

$$=\frac{u^2}{2}\Big]_2^{e+1}$$

$$=\frac{1}{2}\big[(e+1)^2-(2)^2\big]$$

$$=\frac{1}{2}\big[e^2+2e+1-4\big]$$

$$=\frac{1}{2}(e^2+2e-3)=\frac{1}{2}e^2+e-\frac{3}{2}$$

$u=e^x+1$
$du=e^x d^x$
if $x=1,\ u=e+1$
if $x=0,\ u=2$

33. $\displaystyle\int_{0}^{1}\left(x^{1/2}-x^{1/3}\right)dx$

$$=\frac{x^{3/2}}{\frac{3}{2}}-\frac{x^{4/3}}{\frac{4}{3}}\Big]_0^1$$

$$=\frac{2}{3}\sqrt{x^3}-\frac{3}{4}\sqrt[3]{x^4}\Big]_0^1$$

$$=\left(\frac{2}{3}-\frac{3}{4}\right)-(0-0)$$

$$=\frac{-1}{12}$$

35. $\displaystyle\int_{-2}^{1}x(x+3)^{1/2}dx$

$$=\int_{1}^{4}(u-3)u^{1/2}du$$

$$=\int_{1}^{4}\left(u^{3/2}-3u^{1/2}\right)du$$

$$=\frac{u^{5/2}}{\frac{5}{2}}-3\frac{u^{3/2}}{\frac{3}{2}}\Big]_1^4$$

$u=x+3$
$du=dx$
$u-3=x$
if $x=1,\ u=4$
if $x=-2,\ u=1$

$$=\frac{2}{5}\sqrt{u^5} - 2\sqrt{u^3}\Big]_1^4$$

$$=\left[\frac{2}{5}(32) - 2(8)\right] - \left[\frac{2}{5}(1) - 2(1)\right]$$

$$=\frac{64}{5} - 16 - \frac{2}{5} + 2$$

$$=\frac{-8}{5}$$

37.

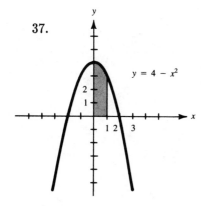

$$\int_0^1 (4 - x^2)\,dx$$

$$=4x - \frac{x^3}{3}\Big]_0^1$$

$$=\left(4 - \frac{1}{3}\right) - 0$$

$$=\frac{11}{3}$$

39.

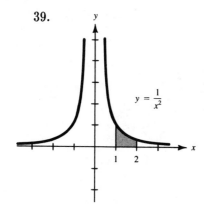

$$\int_1^2 x^{-2}\,dx$$

$$=\frac{x^{-1}}{-1}\Big]_1^2$$

$$=\frac{-1}{x}\Big]_1^2$$

$$=\frac{-1}{2} - \left(\frac{-1}{1}\right)$$

$$=\frac{1}{2}$$

41.

$y = 6 - x - x^2$

$$6 - x - x^2 = 0$$

$$(3 + x)(2 - x) = 0$$

$$x = -3 \qquad x = 2$$

$$\int_{-3}^2 (6 - x - x^2)\,dx$$

$$=6x - \frac{x^2}{2} - \frac{x^3}{3}\Big]_{-3}^2$$

$$=\left[12 - 2 - \frac{8}{3}\right] - \left[-18 - \frac{9}{2} - (-9)\right]$$

$$=\frac{22}{3} - \left(\frac{-27}{2}\right) = \frac{125}{6}$$

43.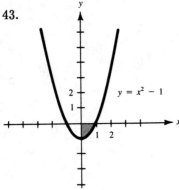

$$y = x^2 - 1$$

$$\left| \int_0^1 (x^2 - 1)dx \right|$$

$$= \left| \frac{x^3}{3} - x \bigg]_0^1 \right|$$

$$= \left| \left(\frac{1}{3} - 1 \right) - 0 \right|$$

$$= \left| \frac{-2}{3} \right|$$

$$= \frac{2}{3}$$

45.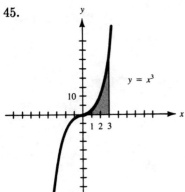

$$y = x^3$$

$$\int_1^3 x^3 dx$$

$$= \frac{x^4}{4} \bigg]_1^3$$

$$= \frac{81}{4} - \frac{1}{4}$$

$$= 20$$

47.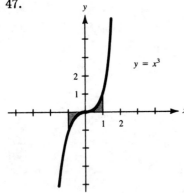

$$y = x^3$$

$$\left| \int_{-1}^0 x^3 dx \right| + \int_0^1 x^3 dx$$

$$= \left| \frac{x^4}{4} \bigg]_{-1}^0 \right| + \frac{x^4}{4} \bigg]_0^1$$

$$= \left| 0 - \frac{1}{4} \right| + \left(\frac{1}{4} - 0 \right)$$

$$= \frac{1}{4} + \frac{1}{4}$$

$$= \frac{1}{2}$$

49.

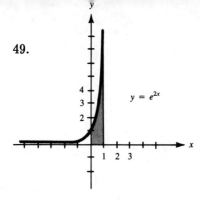

$y = e^{2x}$

$$\int_0^1 e^{2x}dx$$

$$=\frac{1}{2}\int_0^1 e^{2x}2dx$$

$$=\frac{1}{2}\int_0^2 e^u du$$

$$=\frac{1}{2}e^u \Big|_0^2$$

$$=\frac{1}{2}(e^2 - e^0)$$

$$=\frac{1}{2}(e^2 - 1) \approx 3.19$$

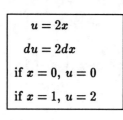

$u = 2x$
$du = 2dx$
if $x = 0$, $u = 0$
if $x = 1$, $u = 2$

51.

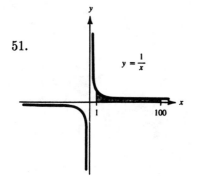

$y = \frac{1}{x}$

$$\int_1^{100} \frac{1}{x}dx$$

$$=\ln |x| \Big|_1^{100}$$

$$=\ln 100 - \ln 1$$

$$\approx 4.61$$

53.

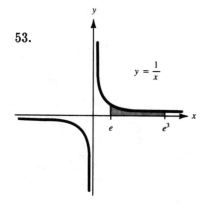

$y = \frac{1}{x}$

$$\int_e^{e^3} \frac{1}{x}dx$$

$$=\ln |x| \Big|_e^{e^3}$$

$$=\ln e^3 - \ln e$$

$$=3 - 1 = 2$$

55. $\displaystyle\int_{-1}^{1}(x^2+1)dx+\int_{1}^{3}(3-x)dx$

$\displaystyle=\left(\frac{x^3}{3}+x\right)\Big]_{-1}^{1}+\left(3x-\frac{x^2}{2}\right)\Big]_{1}^{3}$

$\displaystyle=\left(\tfrac{1}{3}+1\right)-\left(\tfrac{-1}{3}-1\right)+\left(9-\tfrac{9}{2}\right)-\left(3-\tfrac{1}{2}\right)$

$\displaystyle=\tfrac{4}{3}+\tfrac{4}{3}+\tfrac{9}{2}-\tfrac{5}{2}=\tfrac{14}{3}$

57. $\frac{1}{2}\ln 1.5+\frac{1}{2}\ln 2+\frac{1}{2}\ln 2.5+\frac{1}{2}\ln 3+\frac{1}{2}\ln 3.5+\frac{1}{2}\ln 4+\frac{1}{2}\ln 4.5+\frac{1}{2}\ln 5$

$=\frac{1}{2}(\ln 1.5+\ln 2+\ln 2.5+\ln 3+\ln 3.5+\ln 4+\ln 4.5+\ln 5)$

≈ 4.433044

59. $\displaystyle\int_{10}^{16}(100+200x-12x^2)dx$

$=100x+100x^2-4x^3\big]_{10}^{16}$

$=(1600+25600-16384)-(1000+10000-4000)$

$=\$3816$

61. $\displaystyle\int_{3}^{8}(8x+3)dx$

$=4x^2+3x\big]_{3}^{8}$

$=(256+24)-(36+9)$

$=235 \qquad \$235,000$

63. $\displaystyle\int_{0}^{10}10e^{0.1t}dt$

$\displaystyle(10)(10)\int_{0}^{10}e^{0.1t}(0.1dt)$

$\displaystyle100\int_{0}^{1}e^{u}du$

$=100e^{u}\big]_{0}^{1}$

$=100(e^1-e^0)$

$=100(e-1)$

≈ 172

| $u=0.1t$ |
| $du=0.1dt$ |
| if $t=0$, $u=0$ |
| if $t=10$, $u=1$ |

65. $\displaystyle\int_0^5(-32t+108)dt$

$=-16t^2+108t\Big]_0^5$

$=(-400+540)-0$

$=140$ ft.

Section 6.4 (pages 409-412)

1. $\displaystyle\int_0^4[3x-x]dx$

$=\displaystyle\int_0^4 2x\,dx$

$=x^2\Big]_0^4$

$=16-0=16$

3. $\displaystyle\int_{-1}^1[(4-x^2)-x^2]dx$

$=\displaystyle\int_{-1}^1(4-2x^2)dx$

$=4x-\dfrac{2}{3}x^3\Big]_{-1}^1$

$=\left(4-\dfrac{2}{3}\right)-\left(-4+\dfrac{2}{3}\right)$

$=\dfrac{20}{3}$

5. $\displaystyle\int_0^2[(x^3+2)-(-x^2)]dx$

$=\displaystyle\int_0^2(x^3+x^2+2)dx$

$=\dfrac{x^4}{4}+\dfrac{x^3}{3}+2x\Big]_0^2$

$=\left(4+\dfrac{8}{3}+4\right)-0$

$=\dfrac{32}{3}$

7. $\displaystyle\int_1^2[(1-x^2)-(x^2-4x+1)]dx$

$=\displaystyle\int_1^2(-2x^2+4x)dx$

$=\dfrac{-2}{3}x^3+2x^2\Big]_1^2$

$=\left(\dfrac{-16}{3}+8\right)-\left(\dfrac{-2}{3}+2\right)$

$=\dfrac{4}{3}$

9.

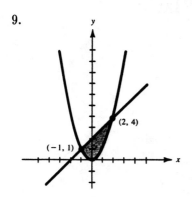

$x^2=x+2$

$x^2-x-2=0$

$(x-2)(x+1)=0$

$x=2 \qquad x=-1$

$\displaystyle\int_{-1}^2[(x+2)-x^2]dx$

$\displaystyle\int_{-1}^2(2+x-x^2)dx$

$=2x+\dfrac{x^2}{2}-\dfrac{x^3}{3}\Big]_{-1}^2$

$=\left(4+2-\dfrac{8}{3}\right)-\left(-2+\dfrac{1}{2}+\dfrac{1}{3}\right)$

$=\dfrac{9}{2}$

11.

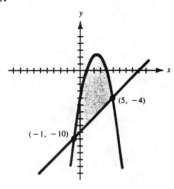

$$-x^2 + 5x - 4 = x - 9$$
$$0 = x^2 - 4x - 5$$
$$0 = (x-5)(x+1)$$
$$x = 5 \qquad x = -1$$

$$\int_{-1}^{5} \left[(-x^2 + 5x - 4) - (x - 9) \right] dx$$
$$= \int_{-1}^{5} (-x^2 + 4x + 5) dx$$
$$= \frac{-x^3}{3} + 2x^2 + 5x \Big]_{-1}^{5}$$
$$= \left(\frac{-125}{3} + 50 + 25 \right) - \left(\frac{1}{3} + 2 - 5 \right)$$
$$= 36$$

13.

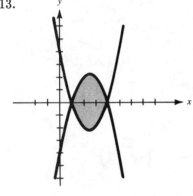

$$x^2 - 5x + 4 = -x^2 + 5x - 4$$
$$2x^2 - 10x + 8 = 0$$
$$2(x^2 - 5x + 4) = 0$$
$$2(x-1)(x-4) = 0$$
$$x = 1 \qquad x = 4$$

$$\int_{1}^{4} \left[(-x^2 + 5x - 4) - (x^2 - 5x + 4) \right] dx$$
$$= \int_{1}^{4} (-2x^2 + 10x - 8) dx$$
$$= \frac{-2}{3} x^3 + 5x^2 - 8x \Big]_{1}^{4}$$
$$= \left(\frac{-128}{3} + 80 - 32 \right) - \left(\frac{-2}{3} + 5 - 8 \right)$$
$$= 9$$

15.

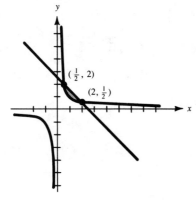

$(\frac{1}{2}, 2)$

$(2, \frac{1}{2})$

$$\frac{1}{x} = -x + \frac{5}{2}$$

$$2 = -2x^2 + 5x$$

$$2x^2 - 5x + 2 = 0$$

$$(2x - 1)(x - 2) = 0$$

$$x = \frac{1}{2} \qquad x = 2$$

$$\int_{\frac{1}{2}}^{2} \left[\left(-x + \frac{5}{2}\right) - \frac{1}{x}\right] dx$$

$$= \frac{-x^2}{2} + \frac{5}{2}x - \ln|x| \Big]_{\frac{1}{2}}^{2}$$

$$= \left(-2 + 5 - \ln 2\right) - \left(\frac{-1}{8} + \frac{5}{4} - \ln \frac{1}{2}\right)$$

$$\approx 0.49$$

17. $\displaystyle\int_{1}^{2} [0 - (-x^3)] dx$

$$= \int_{1}^{2} x^3 dx$$

$$= \frac{x^4}{4} \Big]_{1}^{2}$$

$$= 4 - \frac{1}{4} = \frac{15}{4}$$

19. $\displaystyle\int_{-2}^{0} \left[\frac{1}{2}x^3 - 2x\right] dx + \int_{0}^{2} \left[2x - \frac{1}{2}x^3\right] dx$

$$= \left(\frac{1}{8}x^4 - x^2\right)\Big]_{-2}^{0} + \left(x^2 - \frac{1}{8}x^4\right)\Big]_{0}^{2}$$

$$= 0 - (2 - 4) + (4 - 2) - 0$$

$$= 4$$

21. $x^3 - 3x = 2x^2$

$$x^3 - 2x^2 - 3x = 0$$

$$x(x^2 - 2x - 3) = 0$$

$$x(x + 1)(x - 3) = 0$$

$$x = 0 \qquad x = -1 \qquad x = 3$$

$$\int_{-1}^{0} \left[(x^3 - 3x) - 2x^2\right] dx + \int_{0}^{3} \left[2x^2 - (x^3 - 3x)\right] dx$$

$$= \left(\frac{x^4}{4} - \frac{3}{2}x^2 - \frac{2}{3}x^3\right)\Big]_{-1}^{0} + \left(\frac{2}{3}x^3 - \frac{x^4}{4} + \frac{3}{2}x^2\right)\Big]_{0}^{3}$$

$$= 0 - \left(\frac{1}{4} - \frac{3}{2} + \frac{2}{3}\right) + \left(18 - \frac{81}{4} + \frac{27}{2}\right) - 0$$

$$= \frac{71}{6}$$

23.

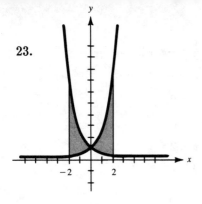

$$\int_{-2}^{0}[e^{-x}-e^{x}]dx + \int_{0}^{2}[e^{x}-e^{-x}]dx$$

$$=(-e^{-x}-e^{x})\Big]_{-2}^{0}+(e^{x}+e^{-x})\Big]_{0}^{2}$$

$$=(-e^{0}-e^{0})-(-e^{2}-e^{-2})+(e^{2}+e^{-2})-(e^{0}+e^{0})$$

$$=-4+2e^{2}+2e^{-2}$$

$$\approx 11.05$$

25a.
$$\int_{0}^{2}(x+2)dx$$

$$=\frac{x^2}{2}+2x\Big]_{0}^{2}$$

$$=(2+4)-0=6$$

25b.
$$\left|\int_{0}^{2}(-x^2)dx\right|$$

$$=\left|\frac{-x^3}{3}\Big]_{0}^{2}\right|$$

$$=\left|\frac{-8}{3}\right|=\frac{8}{3}$$

25c. $6+\frac{8}{3}=\frac{26}{3}$

25d. They are the same.

27.

$$50-x=0.05x^2+10$$

$$0=0.05x^2+x-40$$

$$0=x^2+20x-800$$

$$0=(x+40)(x-20)$$

$$x=-40 \qquad x=20$$

if $x=20$, $p=50-20=30$

equilibrium point $(20,30)$

$$\text{consumer surplus}=\int_{0}^{20}[(50-x)-30]dx$$

$$=\int_{0}^{20}(20-x)dx$$

$$=20x-\frac{1}{2}x^2\Big]_{0}^{20}$$

$$=400-200$$

$$=\$200$$

producer surplus

$$=\int_{0}^{20}[30-(0.05x^2+10)]dx$$

$$=\int_{0}^{20}[20-0.05x^2]dx$$

$$=20x-\frac{x^3}{60}\Big]_{0}^{20}$$

$$=400-\frac{400}{3}$$

$$\approx \$266.67$$

29.

$$500 - 0.2x^2 = 48x$$

$$-0.2x^2 - 48x + 500 = 0$$

$$x^2 + 240x - 2500 = 0$$

$$(x + 250)(x - 10) = 0$$

$$x = -250 \qquad x = 10$$

if $x = 10$, $p = 48(10) = 480$

equilibrium point $(10, 480)$

consumer surplus $= \displaystyle\int_0^{10} [(500 - 0.2x^2) - 480]dx$

$$= \int_0^{10} (20 - 0.2x^2)dx$$

$$= 20x - \frac{2}{30}x^3 \Big|_0^{10}$$

$$= 200 - \frac{200}{3}$$

$$\approx \$133.33$$

producer surplus $= \displaystyle\int_0^{10} [480 - 48x]dx$

$$= 480x - 24x^2 \Big|_0^{10}$$

$$= 4800 - 2400$$

$$= \$2400$$

31.

$$4750 - x^2 = 0.5x^2 + 1000$$

$$-1.5x^2 = -3750$$

$$x^2 = 2500$$

$$x = \pm 50$$

$$x^* = 50$$

33.

$$3600 - x^2 = 0.5x^2 + 1200$$

$$-1.5x^2 = -2400$$

$$x^2 = 1600$$

$$x = \pm 40$$

$$x^* = 40$$

35.

$$3000 - x^2 = 2x + 600$$

$$0 = x^2 + 2x - 2400$$

$$0 = (x - 48)(x + 50)$$

$$x = 48 \qquad x = -50$$

$$x^* = 48$$

37. $p = 3e^{3/3} = 3e$

consumer surplus

$$= \int_0^3 [3e - 3e^{x/3}]dx$$

$$= 3ex - 9e^{x/3} \Big|_0^3$$

$$= \left[3e(3) - 9e^{3/3}\right] - \left[0 - 9e^0\right]$$

$$= 9e - 9e + 9$$

$$= 9$$

39.

$$\int_5^8 [A(t) - B(t)]dt = \int_5^8 \left[300(t+4)^{1/2} - (1150 - 50t)\right]dt$$

$$= \int_5^8 \left[300(t+4)^{1/2} - 1150 + 50t\right]dt$$

$$= 300\frac{(t+4)^{3/2}}{\frac{3}{2}} - 1150t + 25t^2 \Big]_5^8$$

$$= 200\sqrt{(t+4)^3} - 1150t + 25t^2 \Big]_5^8$$

$$= [200\sqrt{12^3} - 9200 + 1600] - [5400 - 5750 + 625]$$

$$\approx \$438{,}844$$

Section 6.5   (pages 418-420)

1.
$$f_{av} = \frac{1}{2-0}\int_0^2 2x\,dx$$
$$= \frac{1}{2}[x^2]_0^2$$
$$= \frac{1}{2}[4-0]$$
$$= 2$$

3.
$$f_{av} = \frac{1}{1-0}\int_0^1 (x^3+1)dx$$
$$= 1\left[\frac{x^4}{4} + x\right]_0^1$$
$$= \left(\frac{1}{4}+1\right) - 0$$
$$= \frac{5}{4}$$

5.
$$y_{av} = \frac{1}{1-(-1)}\int_{-1}^1 (1-x^2)dx$$
$$= \frac{1}{2}\left[x - \frac{x^3}{3}\right]_{-1}^1$$
$$= \frac{1}{2}\left[\left(1-\frac{1}{3}\right) - \left(-1+\frac{1}{3}\right)\right]$$
$$= \frac{1}{2}\left(\frac{4}{3}\right) = \frac{2}{3}$$

7.
$$f_{av} = \frac{1}{4-0}\int_0^4 x^{1/2}dx$$
$$= \frac{1}{4}\left[\frac{x^{3/2}}{\frac{3}{2}}\right]_0^4$$
$$= \frac{1}{6}\sqrt{x^3}\Big]_0^4$$
$$= \frac{4}{3}$$

9.
$$\text{volume} = \int_0^1 \pi(x+3)^2 dx$$
$$= \pi\int_0^1 (x^2 + 6x + 9)dx$$
$$= \pi\left[\frac{x^3}{3} + 3x^2 + 9x\right]_0^1$$
$$= \pi\left[\frac{1}{3} + 3 + 9\right]$$
$$= \frac{37\pi}{3}$$

11.
$$\text{volume} = \int_{-1}^1 \pi(1-x^2)^2 dx$$
$$= \pi\int_{-1}^1 (1 - 2x^2 + x^4)dx$$
$$= \pi\left[x - \frac{2}{3}x^3 + \frac{x^5}{5}\right]_{-1}^1$$
$$= \pi\left[\left(1 - \frac{2}{3} + \frac{1}{5}\right) - \left(-1 + \frac{2}{3} - \frac{1}{5}\right)\right]$$
$$= \frac{16\pi}{15}$$

13.

$$\text{volume} = \int_0^1 \pi(x^3)^2 dx$$

$$= \pi \int_0^1 x^6 dx$$

$$= \pi \left[\frac{x^7}{7}\right]_0^1$$

$$= \pi \left[\frac{1}{7} - 0\right]$$

$$= \frac{\pi}{7}$$

15.

$$\text{volume} = \int_0^3 \pi(\sqrt{x+1})^2 dx$$

$$= \pi \int_0^3 (x+1) dx$$

$$= \pi \left[\frac{x^2}{2} + x\right]_0^3$$

$$= \pi \left[\frac{9}{2} + 3\right]$$

$$= \frac{15\pi}{2}$$

17.

$$\text{volume} = \int_0^1 \pi(e^x)^2 dx$$

$$= \pi \int_0^1 e^{2x} dx$$

$$= \pi \left[\frac{e^{2x}}{2}\right]_0^1$$

$$= \pi \left[\frac{1}{2}e^2 - \frac{1}{2}\right]$$

$$\approx 10.036$$

19.

$$\text{volume} = \int_1^2 \pi(x^{-1/2})^2 dx$$

$$= \pi \int_1^2 x^{-1} dx$$

$$= \pi \left[\ln|x|\right]_1^2$$

$$= \pi[\ln 2 - 0]$$

$$\approx 2.178$$

21.

$$\text{volume} = \int_0^4 \pi(2)^2 dx$$

$$= 4\pi \int_0^4 dx$$

$$= 4\pi [x]_0^4$$

$$= 4\pi(4)$$

$$= 16\pi$$

23.

$$2x - x^2 = x$$

$$x - x^2 = 0$$

$$x(1-x) = 0$$

$$x = 0 \qquad x = 1$$

(This is similiar to the process used to find the area between two curves.)

$$\text{volume} = \pi \int_0^1 \left([f(x)]^2 - [g(x)]^2\right) dx$$

$$= \pi \int_0^1 \left[(2x - x^2)^2 - x^2\right] dx$$

$$= \pi \int_0^1 (4x^2 - 4x^3 + x^4 - x^2) dx$$

$$= \pi \int_0^1 (3x^2 - 4x^3 + x^4) dx$$

$$= \pi \left[x^3 - x^4 + \frac{x^5}{5}\right]_0^1$$

$$= \pi \left[1 - 1 + \frac{1}{5}\right]$$

$$= \frac{\pi}{5}$$

25. $C_{av} = \dfrac{1}{10-0}\displaystyle\int_0^{10}(1.5t^2+0.5)dt$

$= \dfrac{1}{10}\Big[0.5t^3+0.5t\Big]_0^{10}$

$= \dfrac{1}{10}[500+5]$

$= 50.5 \qquad$ average cost: $50,500

27. $y_{av} = \dfrac{1}{10-0}\displaystyle\int_0^{10}(600-600e^{-0.3t})dt$

$= \dfrac{1}{10}\Big[600t+2000e^{-0.3t}\Big]_0^{10}$

$= \dfrac{1}{10}\Big[(6000+2000e^{-3})-(0+2000)\Big]$

≈ 410

29. $P_{av} = \dfrac{1}{10-5}\displaystyle\int_5^{10}(e^{-0.2x}+5)dx$

$= \dfrac{1}{5}\Big[-5e^{-0.2x}+5x\Big]_5^{10}$

$= \dfrac{1}{5}\Big[(-5e^{-2}+50)-(-5e^{-1}+25)\Big]$

≈ 5.233

31. $S_{av} = \dfrac{1}{10-6}\displaystyle\int_6^{10}(90+4.8t-0.08t^2)dt$

$= \dfrac{1}{4}\Big[90t+2.4t^2-\dfrac{8}{300}t^3\Big]_6^{10}$

$= \dfrac{1}{4}\Big[(900+240-\dfrac{80}{3})-(540+86.4-\dfrac{144}{25})\Big]$

≈ 123.2

Chapter 6 Review   (pages 421-423)

1. $\displaystyle\int(3x^2-x+1)dx$

$=x^3-\dfrac{x^2}{2}+x+C$

2. $\displaystyle\int(4x^3-4x)dx$

$=x^4-2x^2+C$

3. $\displaystyle\int x^2(1-x^3)^{1/2}dx$

$\boxed{\begin{array}{l} u=-x^3 \\[4pt] du=-3x^2dx \end{array}}$

$=\dfrac{-1}{3}\displaystyle\int (1-x^3)^{1/2}(-3x^2dx)$

$=\dfrac{-1}{3}\displaystyle\int u^{1/2}du$

$=\dfrac{-1}{3}\dfrac{u^{3/2}}{\frac{3}{2}}+C$

$=\dfrac{-2}{9}(1-x^3)^{3/2}+C$

4. $\displaystyle\int (1-x^3)^{-1/2}x^2dx$

$\boxed{\begin{array}{l} u=-x^3 \\[4pt] du=-3x^2dx \end{array}}$

$=\dfrac{-1}{3}\displaystyle\int (1-x^3)^{-1/2}(-3x^2dx)$

$=\dfrac{-1}{3}\displaystyle\int u^{-1/2}du$

$=\dfrac{-1}{3}\dfrac{u^{1/2}}{\frac{1}{2}}+C$

$=\dfrac{-2}{3}\sqrt{1-x^3}+C$

5. $\displaystyle\int (2x-1)^5dx$

$\boxed{\begin{array}{l} u=2x-1 \\[4pt] du=2dx \end{array}}$

$=\dfrac{1}{2}\displaystyle\int (2x-1)^5(2dx)$

$=\dfrac{1}{2}\displaystyle\int u^5du$

$=\dfrac{1}{2}\dfrac{u^6}{6}+C$

$=\dfrac{1}{12}(2x-1)^6+C$

6. $\displaystyle\int (2x-1)^{1/5}dx$

$\boxed{\begin{array}{l} u=2x-1 \\[4pt] du=2dx \end{array}}$

$=\dfrac{1}{2}\displaystyle\int (2x-1)^{1/5}(2dx)$

$=\dfrac{1}{2}\displaystyle\int u^{1/5}du$

$=\dfrac{1}{2}\dfrac{u^{6/5}}{\frac{6}{5}}+C$

$=\dfrac{5}{12}(2x-1)^{6/5}+C$

7. $\int \dfrac{1}{2x+5}dx$

$=\dfrac{1}{2}\int \dfrac{1}{2x+5}(2dx)$

$=\dfrac{1}{2}\int \dfrac{1}{u}du$

$=\dfrac{1}{2}\ln|u|+C$

$=\dfrac{1}{2}\ln|2x+5|+C$

$\boxed{\begin{array}{l} u=2x+5 \\ du=2dx \end{array}}$

8. $\int \dfrac{1}{2x^2+5}dx$

$=\dfrac{1}{4}\int \dfrac{1}{2x^2+5}(4xdx)$

$=\dfrac{1}{4}\int \dfrac{1}{u}du$

$=\dfrac{1}{4}\ln|u|+C$

$=\dfrac{1}{4}\ln(2x^2+5)+C$

$\boxed{\begin{array}{l} u=2x^2+5 \\ du=4xdx \end{array}}$

9. $\int \dfrac{1}{x\sqrt{x}}dx$

$=\int \dfrac{1}{x^{3/2}}dx$

$=\int x^{-3/2}dx$

$=\dfrac{x^{-1/2}}{\frac{-1}{2}}+C$

$=\dfrac{-2}{\sqrt{x}}+C$

10. $\int \dfrac{x^2-1}{x^{1/2}}dx$

$=\int \left(\dfrac{x^2}{x^{1/2}}-\dfrac{1}{x^{1/2}}\right)dx$

$=\int \left(x^{3/2}-x^{-1/2}\right)dx$

$=\dfrac{x^{5/2}}{\frac{5}{2}}-\dfrac{x^{1/2}}{\frac{1}{2}}+C$

$=\dfrac{2}{5}x^{5/2}-2x^{1/2}+C$

11. $\int_{0}^{1}(x^2-x+3)dx$

$=\dfrac{x^3}{3}-\dfrac{x^2}{2}+3x\Big]_{0}^{1}$

$=\left(\dfrac{1}{3}-\dfrac{1}{2}+3\right)-0$

$=\dfrac{17}{6}$

12. $\int_{1}^{8}2x^{-1/3}dx$

$=2\dfrac{x^{2/3}}{\frac{2}{3}}\Big]_{1}^{8}$

$=3\sqrt[3]{x^2}\Big]_{1}^{8}$

$=3(4)-3(1)=9$

13. $\displaystyle\int_0^2 (x-2)^4 dx$

$= \displaystyle\int_{-2}^0 u^4 du$

$= \dfrac{u^5}{5}\Big]_{-2}^0$

$= 0 - \dfrac{-32}{5} = \dfrac{32}{5}$

$u = x - 2$
$du = dx$
if $x = 0$, $u = -2$
if $x = 2$, $u = 0$

14. $\displaystyle\int_2^4 (x-2)^4 dx$

$= \displaystyle\int_0^2 u^4 du$

$= \dfrac{u^5}{5}\Big]_0^2$

$= \dfrac{32}{5} - 0 = \dfrac{32}{5}$

$u = x - 2$
$du = dx$
if $x = 2$, $u = 0$
if $x = 4$, $u = 2$

15. $\displaystyle\int_0^1 e^{4x} dx$

$= \dfrac{1}{4}\displaystyle\int_0^1 e^{4x}(4dx)$

$= \dfrac{1}{4}\displaystyle\int_0^4 e^u du$

$= \dfrac{1}{4} e^u \Big]_0^4$

$= \dfrac{1}{4} e^4 - \dfrac{1}{4} \approx 13.400$

$u = 4x$
$du = 4dx$
if $x = 0$, $u = 0$
if $x = 1$, $u = 4$

16. $\displaystyle\int_1^2 \dfrac{3}{(2x-1)^2} dx$

$= 3 \cdot \dfrac{1}{2}\displaystyle\int_1^2 (2x-1)^{-2}(2dx)$

$= \dfrac{3}{2}\displaystyle\int_1^3 u^{-2} du$

$= \dfrac{3}{2} \dfrac{u^{-1}}{-1}\Big]_1^3$

$= \dfrac{-3}{2u}\Big]_1^3$

$= \dfrac{-1}{2} - \left(\dfrac{-3}{2}\right) = 1$

$u = 2x - 1$
$du = 2dx$
if $x = 1$, $u = 1$
if $x = 2$, $u = 3$

17. $\displaystyle\int_{-1}^{1}(x+3)^{-1/2}dx$

$\boxed{\begin{aligned} u &= x+3 \\ du &= dx \\ &\text{if } x=-1,\ u=2 \\ &\text{if } x=1,\ u=4 \end{aligned}}$

$\displaystyle=\int_{2}^{4}u^{-1/2}du$

$\displaystyle=\frac{u^{1/2}}{\frac{1}{2}}\Big]_{2}^{4}$

$\displaystyle=2\sqrt{u}\,\Big]_{2}^{4}$

$\displaystyle=2\sqrt{4}-2\sqrt{2}=4-2\sqrt{2}$

18. $\displaystyle\int_{-2}^{-1}x^{-2}dx$

$\displaystyle=\frac{x^{-1}}{-1}\Big]_{-2}^{-1}$

$\displaystyle=\frac{-1}{x}\Big]_{-2}^{-1}$

$\displaystyle=1-\frac{1}{2}=\frac{1}{2}$

19. $\displaystyle\int_{-1}^{1}(1-x^{4})dx$

$\displaystyle=x-\frac{x^{5}}{5}\Big]_{-1}^{1}$

$\displaystyle=(1-\tfrac{1}{5})-(-1+\tfrac{1}{5})$

$\displaystyle=\frac{8}{5}$

20. $\displaystyle\int_{-1}^{1}(x^{3}+2)dx$

$\displaystyle=\frac{x^{4}}{4}+2x\,\Big]_{-1}^{1}$

$\displaystyle=(\tfrac{1}{4}+2)-(\tfrac{1}{4}-2)=4$

21. $\displaystyle\int_{0}^{3}(2x+1)dx$

$\displaystyle=x^{2}+x\,\Big]_{0}^{3}$

$\displaystyle=(9+3)-0=12$

22. $\displaystyle\int_{0}^{1}(1-x)dx$

$\displaystyle=x-\frac{x^{2}}{2}\Big]_{0}^{1}$

$\displaystyle=(1-\tfrac{1}{2})-0=\frac{1}{2}$

23. $\displaystyle\int_{0}^{1}(x^{2}-x^{3})dx$

$\displaystyle=\frac{x^{3}}{3}-\frac{x^{4}}{4}\Big]_{0}^{1}$

$\displaystyle=(\tfrac{1}{3}-\tfrac{1}{4})-0=\frac{1}{12}$

24. $\qquad x=2x-x^{2}$

$\qquad x^{2}-x=0$

$\qquad x(x-1)=0$

$\qquad x=0 \qquad x=1$

$\displaystyle\int_{0}^{1}\big[(2x-x^{2})-x\big]dx$

$\displaystyle\int_{0}^{1}(x-x^{2})dx$

$\displaystyle=\frac{x^{2}}{2}-\frac{x^{3}}{3}\Big]_{0}^{1}$

$\displaystyle=\frac{1}{2}-\frac{1}{3}=\frac{1}{6}$

25. $f_{av} = \dfrac{1}{6-2}\displaystyle\int_2^6 (x-2)^{1/2}dx$

$$\boxed{\begin{array}{l} u = x - 2 \\[4pt] du = dx \\[4pt] \text{if } x = 2,\ u = 0 \\[4pt] \text{if } x = 6,\ u = 4 \end{array}}$$

$\quad = \dfrac{1}{4}\displaystyle\int_0^4 u^{1/2}du$

$\quad = \dfrac{1}{4}\dfrac{u^{3/2}}{\frac{3}{2}}\Big]_0^4$

$\quad = \dfrac{1}{6}\sqrt{u^3}\Big]_0^4$

$\quad = \dfrac{1}{6}[8-0] = \dfrac{4}{3}$

26. $f_{av} = \dfrac{1}{4-1}\displaystyle\int_1^4 x^{-2}dx$

$\quad = \dfrac{1}{3}\dfrac{x^{-1}}{-1}\Big]_1^4$

$\quad = \dfrac{-1}{3x}\Big]_1^4$

$\quad = \left(\dfrac{-1}{12}\right) - \left(\dfrac{-1}{3}\right) = \dfrac{1}{4}$

27. $\text{volume} = \displaystyle\int_1^9 \pi(\sqrt{x})^2 dx$

$\quad = \pi\displaystyle\int_1^9 x\,dx$

$\quad = \pi\dfrac{x^2}{2}\Big]_1^9$

$\quad = \pi\left[\dfrac{81}{2} - \dfrac{1}{2}\right] = 40\pi$

28. $\text{volume} = \displaystyle\int_0^1 \pi(x^{2/3})^2 dx$

$\quad = \pi\displaystyle\int_0^1 x^{4/3} dx$

$\quad = \pi\left[\dfrac{x^{7/3}}{\frac{7}{3}}\right]_0^1$

$\quad = \pi\left[\dfrac{3}{7}\sqrt[3]{x^7}\right]_0^1$

$\quad = \pi\left[\dfrac{3}{7} - 0\right]$

$\quad = \dfrac{3\pi}{7}$

29. $f(x) = \displaystyle\int f'(x)dx$

$\quad = \displaystyle\int (x^2 - x^{-2})dx$

$\quad = \dfrac{x^3}{3} - \dfrac{x^{-1}}{-1} + C$

$\quad = \dfrac{x^3}{3} + \dfrac{1}{x} + C$

$\quad f(1) = \dfrac{1}{3} + 1 + C$

$\quad \dfrac{4}{3} = \dfrac{4}{3} + C$

$\quad 0 = C$

$\quad f(x) = \dfrac{x^3}{3} + \dfrac{1}{x}$

30. $f(x) = \int f'(x)dx$

$\boxed{\begin{aligned} u &= 1 - 2x \\ du &= -2dx \end{aligned}}$

$\quad = \int (1 - 2x)^{1/2}dx$

$\quad = \frac{-1}{2}\int (1 - 2x)^{1/2}(-2dx)$

$\quad = \frac{-1}{2}\int u^{1/2}du$

$\quad = \frac{-1}{2}\frac{u^{3/2}}{\frac{3}{2}} + C$

$\quad = \frac{-1}{3}\sqrt{u^3} + C$

$\quad = \frac{-1}{3}\sqrt{(1 - 2x)^3} + C$

$f(-4) = \frac{-1}{3}\left(\sqrt{1 - 2(-4)}\right)^3 + C$

$-3 = -9 + C$

$6 = C$

$f(x) = \frac{-1}{3}\sqrt{(1 - 2x)^3} + 6$

31. $P(x) = \int P'(x)dx$

$\boxed{\begin{aligned} u &= x - 5 \\ du &= dx \end{aligned}}$

$\quad = \int 200(x - 5)^{-2/3}dx$

$\quad = 200\int u^{-2/3}du$

$\quad = 200\frac{u^{1/3}}{\frac{1}{3}} + C$

$\quad = 600\sqrt[3]{x - 5} + C$

$P(5) = 600\sqrt[3]{5 - 5} + C$

$0 = 0 + C$

$0 = C$

$P(x) = 600\sqrt[3]{x - 5}$

32. $700 - 20x = 100 + 0.5x^2$

$\quad\quad 0 = 0.5x^2 + 20x - 600$

$\quad\quad 0 = x^2 + 40x - 1200$

$\quad\quad 0 = (x + 60)(x - 20) \quad\quad x = -60, \ x = 20$

if $x = 20$, $p = 700 - 20(20) = 300$

equilibrium point: $(20, 300)$

consumer surplus $= \int_0^{20} [(700 - 20x) - 300]dx$

$= \int_0^{20} (400 - 20x)dx$

$= 400x - 10x^2 \Big|_0^{20}$

$= (8000 - 4000) - 0 = \$4000$

33. $\int \left(\dfrac{2t}{t^2+1} + 3 \right) dt$

$\boxed{\begin{array}{l} u = t^2+1 \\ du = 2t\,dt \end{array}}$

$= \int \dfrac{2t}{t^2+1}\,dt + \int 3\,dt$

$\int \dfrac{2t}{t^2+1}\,dt = \int \dfrac{1}{u}\,du$

$= \ln|u| + C$

$= \ln|t^2+1| + C$

$\int 3\,dt = 3t + C$

$\dfrac{1}{6-0}\int_0^6 \left(\dfrac{2t}{t^2+1} + 3 \right) dt = \frac{1}{6}\Big[\ln|t^2+1| + 3t\Big]_0^6$

$= \frac{1}{6}[(\ln 37 + 18) - (\ln 1 + 0)] \approx 3.6$

34. producer surplus $= \displaystyle\int_0^{20}[300 - (100 + 0.5x^2)]dx$

$= \displaystyle\int_0^{20}(200 - 0.5x^2)dx$

$= 200x - \frac{1}{6}x^3 \Big]_0^{20}$

$= \left(4000 - \dfrac{4000}{3} \right) - 0 \approx \$2666.67$

35. $C(x) = \displaystyle\int C'(x)\,dx$

$= \displaystyle\int (1 + 10x - e^{-x})\,dx$

$= x + 5x^2 + e^{-x} + C$

$C(0) = 0 + 5(0) + e^0 + C$

$4 = 1 + C \qquad 3 = C$

$C(x) = x + 5x^2 + e^{-x} + 3$

$C(10) = 10 + 500 + e^{-5} + 3 \approx \$513$

Chapter 6 Test (pages 423-425)

1. $f_{av} = \dfrac{1}{3-1}\displaystyle\int_1^3 (x^3 - x)\,dx$

$= \frac{1}{2}\left[\dfrac{x^4}{4} - \dfrac{x^2}{2} \right]_1^3$

$= \frac{1}{2}\left[\left(\dfrac{81}{4} - \dfrac{9}{2} \right) - \left(\dfrac{1}{4} - \dfrac{1}{2} \right) \right]$

$= \frac{1}{2}[16] = 8$ (b)

2. $\displaystyle\int (x^3 - x^{-2})\,dx$

$= \dfrac{x^4}{4} - \dfrac{x^{-1}}{-1} + C$

$= \dfrac{x^4}{4} + \dfrac{1}{x} + C$ (c)

3. $\int e^{-2x}dx$

$u = -2x$

$du = -2dx$

$= \frac{-1}{2}\int e^{-2x}(-2dx)$

$= \frac{-1}{2}\int e^u du$

$= \frac{-1}{2}e^u + C$

$= \frac{-1}{2}e^{-2x} + C$ (d)

4. $\int_0^1 (x - x^2)dx$

$= \frac{x^2}{2} - \frac{x^3}{3}\Big]_0^1$

$= \left(\frac{1}{2} - \frac{1}{3}\right) - 0$

$= \frac{1}{6}$ (b)

5. $\int_0^2 (4x+1)^{1/2}dx$

$= \frac{1}{4}\int_0^2 (4x+1)^{1/2}(4dx)$

$u = 4x + 1$

$du = 4dx$

if $x = 0$, $u = 1$

if $x = 2$, $u = 9$

$= \frac{1}{4}\int_1^9 u^{1/2}du$

$= \frac{1}{4}\frac{u^{3/2}}{\frac{3}{2}}\Big]_1^9$

$= \frac{1}{6}(\sqrt{u})^3\Big]_1^9$

$= \frac{1}{6}[27 - 1] = \frac{13}{3}$ (a)

6. $g(x) = \int g'(x)dx$

$= \int (3x^2 - 2x + 4)dx$

$= x^3 - x^2 + 4x + C$

$g(1) = 1 - 1 + 4 + C$

$3 = 4 + C$

$-1 = C$

$g(x) = x^3 - x^2 + 4x - 1$ (d)

7. c is false. (a, b, and d are given in Theorem 6.1 and Theorem 6.2)

8. $x^2 = 2 - x^2$, $2x^2 = 2$, $x^2 = 1$, $x = \pm 1$

$\int_{-1}^1 [(2 - x^2) - x^2]dx$

$= \int_{-1}^1 (-2x^2 + 2)dx$

$= \frac{-2}{3}x^3 + 2x\Big]_{-1}^1$

$= \left(\frac{-2}{3} + 2\right) - \left(\frac{2}{3} - 2\right) = \frac{8}{3}$ (a)

9. $\text{volume} = \int_0^1 \pi (x^2)^2 \, dx$

$\quad = \pi \int_0^1 x^4 \, dx$

$\quad = \pi \left[\dfrac{x^5}{5} \right]_0^1$

$\quad = \pi \left[\dfrac{1}{5} - 0 \right]$

$\quad = \dfrac{\pi}{5} \quad$ (d)

10. $\int (1-x^2)^{-1/3} 5x \, dx \qquad \boxed{\begin{array}{l} u = 1 - x^2 \\ du = -2x\,dx \end{array}}$

$5\left(\dfrac{-1}{2}\right) \int (1-x^2)^{-1/3}(-2x\,dx)$

$= \dfrac{-5}{2} \int u^{-1/3} \, du$

$= \dfrac{-5}{2} \dfrac{u^{2/3}}{\frac{2}{3}} + C$

$= \dfrac{-15}{4}(1-x^2)^{2/3} + C \qquad$ (b)

11. $\int x^2 (x^3+1)^5 \, dx \qquad \boxed{\begin{array}{l} u = x^3 + 1 \\ du = 3x^2 \, dx \end{array}}$

$= \dfrac{1}{3} \int (x^3+1)^5 (3x^2 \, dx)$

$= \dfrac{1}{3} \int u^5 \, du$

$= \dfrac{1}{3} \dfrac{u^6}{6} + C$

$= \dfrac{1}{18}(x^3+1)^6 + C$

12. $f(x) = \int f'(x) \, dx$

$\quad = \int (2x - x^{-2}) \, dx$

$\quad = x^2 - \dfrac{x^{-1}}{-1} + C$

$\quad = x^2 + \dfrac{1}{x} + C$

$f(1) = 1 + 1 + C$

$2 = 2 + C$

$0 = C$

$f(x) = x^2 + \dfrac{1}{x}$

13. $150 - x = 0.1x^2 + 30$

$0 = 0.1x^2 + x - 120$

$0 = x^2 + 10x - 1200$

$0 = (x+40)(x-30)$

$x = -40, \ x = 30$

if $x = 30$, $p = 150 - 30 = 120$

equilibrium point: $(30, 120)$

$\text{consumer surplus} = \int_0^{30} [(150 - x) - 120] \, dx$

$\quad = \int_0^{30} (30 - x) \, dx$

$\quad = 30x - \dfrac{1}{2}x^2 \Big]_0^{30}$

$\quad = (900 - 450) - 0 = \$450$

$\text{producer surplus} = \int_0^{30} [120 - (0.1x^2 + 30)] \, dx$

$\quad = \int_0^{30} (90 - 0.1x^2) \, dx$

$\quad = 90x - \dfrac{1}{30}x^3 \Big]_0^{30}$

$\quad = (2700 - 900) = \$1800$

14. $\displaystyle\int_5^{10}(2100+0.2x)dx$

$=2100x+0.1x^2\big|_5^{10}$

$=(2100+10)-(10500+2.50)$

$=\$10{,}507.50$

15. $C(x)=\displaystyle\int C'(x)dx$

$=\displaystyle\int(3+4x^2+\tfrac{1}{2}e^{-x})dx$

$=3x+\tfrac{4}{3}x^3-\tfrac{1}{2}e^{-x}+C$

$C(0)=0+0-\tfrac{1}{2}e^0+C$

$5=\tfrac{-1}{2}+C$

$\tfrac{11}{2}=C$

$C(x)=3x+\tfrac{4}{3}x^3-\tfrac{1}{2}e^{-x}+\tfrac{11}{2}$

16. $N_{av}=\dfrac{1}{24-0}\displaystyle\int_0^{24}50e^{0.2t}dt$

$=\tfrac{1}{24}(50)(5)\displaystyle\int_0^{24}e^{0.2t}(0.2dt)$

$=\tfrac{125}{12}\displaystyle\int_0^{4.8}e^u du$

$=\tfrac{125}{12}e^u\big|_0^{4.8}$

$=\tfrac{125}{12}[e^{4.8}-e^0]\approx 1255$

$u=0.2t$
$du=0.2dt$
if $t=0$, $u=0$
if $t=24$, $u=4.8$

CHAPTER 7: ADDITIONAL TOPICS IN INTEGRATION

Section 7.1 (pages 431-433)

1. $\displaystyle\int x^2\ln x\,dx$

$=\ln x\left(\tfrac{1}{3}x^3\right)-\displaystyle\int\tfrac{1}{3}x^3\cdot\tfrac{1}{x}dx$

$=\tfrac{1}{3}x^3\ln x-\tfrac{1}{3}\displaystyle\int x^2 dx$

$=\tfrac{1}{3}x^3\ln x-\tfrac{1}{3}\tfrac{x^3}{3}+C$

$=\tfrac{1}{3}x^3\ln x-\tfrac{1}{9}x^3+C$

$u=\ln x$	$dv=x^2dx$
$du=\tfrac{1}{x}dx$	$v=\tfrac{1}{3}x^3$

3. $\displaystyle\int xe^{x^2}dx$

$=\tfrac{1}{2}\displaystyle\int e^{x^2}(2xdx)$

$=\tfrac{1}{2}\displaystyle\int e^u du$

$=\tfrac{1}{2}e^u+C=\tfrac{1}{2}e^{x^2}+C$

$u=x^2$
$du=2xdx$

5. $\displaystyle\int xe^{-3x}dx$

$u = x$	$dv = e^{-3x}dx$
$du = dx$	$v = \dfrac{-1}{3}e^{-3x}$

$= x\left(\dfrac{-1}{3}e^{-3x}\right) - \displaystyle\int\dfrac{-1}{3}e^{-3x}dx$

$= \dfrac{-1}{3}xe^{-3x} + \dfrac{1}{3}\left(\dfrac{-1}{3}e^{-3x}\right) + C$

$= \dfrac{-1}{3}xe^{-3x} - \dfrac{1}{9}e^{-3x} + C$

7. $\displaystyle\int x^2 e^x dx$

$u = x^2$	$dv = e^x dx$
$du = 2xdx$	$v = e^x$

$= x^2 e^x - \displaystyle\int e^x(2xdx)$

$= x^2 e^x - 2\displaystyle\int e^x xdx$

$u = x$	$dv = e^x dx$
$du = dx$	$v = e^x$

$= x^2 e^x - 2\left[xe^x - \displaystyle\int e^x dx\right]$

$= x^2 e^x - 2[xe^x - e^x] + C$

$= x^2 e^x - 2xe^x + 2e^x + C$

9. $\displaystyle\int \ln 2x\, dx$

$u = \ln 2x$	$dv = dx$
$du = \dfrac{1}{2x}(2)dx$	$v = x$
$= \dfrac{1}{x}dx$	

$= (\ln 2x)(x) - \displaystyle\int x\left(\dfrac{1}{x}dx\right)$

$= x\ln 2x - \displaystyle\int dx$

$= x\ln 2x - x + C$

11. $\displaystyle\int xe^{5x}dx$

$u = x$	$dv = e^{5x}dx$
$du = dx$	$v = \dfrac{1}{5}e^{5x}$

$= x\left(\dfrac{1}{5}e^{5x}\right) - \displaystyle\int\dfrac{1}{5}e^{5x}dx$

$= \dfrac{1}{5}xe^{5x} - \dfrac{1}{5}\cdot\dfrac{1}{5}e^{5x} + C$

$= \dfrac{1}{5}xe^{5x} - \dfrac{1}{25}e^{5x} + C$

13. $\displaystyle\int xe^{-2x}dx$

$u = x$	$dv = e^{-2x}dx$
$du = dx$	$v = \dfrac{-1}{2}e^{-2x}$

$= \left(\dfrac{-1}{2}e^{-2x}\right) - \displaystyle\int\left(\dfrac{-1}{2}e^{-2x}\right)dx$

$= \dfrac{-1}{2}xe^{-2x} + \dfrac{1}{2}\left(\dfrac{-1}{2}e^{-2x}\right) + C$

$= \dfrac{-1}{2}xe^{-2x} - \dfrac{1}{4}e^{-2x} + C$

15. $\int \ln x \left(\frac{1}{x}\right) dx$

$\boxed{\begin{array}{l} u = \ln x \\ du = \frac{1}{x} dx \end{array}}$

$= \int u\, du$

$= \frac{1}{2} u^2 + C$

$= \frac{1}{2} (\ln x)^2 + C$

17. $\int \ln x^{1/2} dx$

$u = \ln x$	$dv = dx$
$du = \frac{1}{x} dx$	$v = x$

$= \frac{1}{2} \int \ln x\, dx$

$= \frac{1}{2} \left[(\ln x)(x) - \int x \left(\frac{1}{x} dx\right) \right]$

$= \frac{1}{2} [x \ln x - \int dx]$

$= \frac{1}{2} [x \ln x - x] + C$

$= \frac{1}{2} x \ln x - \frac{1}{2} x + C$

19. $\int \ln(x^2) dx$

$u = \ln x$	$dv = dx$
$du = \frac{1}{x} dx$	$v = x$

$= 2 \int \ln x\, dx$

$= 2 \left[(\ln x)(x) - \int x \left(\frac{1}{x} dx\right) \right]$

$= 2 [x \ln x - \int dx]$

$= 2 [x \ln x - x] + C$

$= 2x \ln x - 2x + C$

21. $2 \int x \ln x\, dx$

$u = \ln x$	$dv = x dx$
$du = \frac{1}{x} dx$	$v = \frac{1}{2} x^2$

$= 2 \left[(\ln x)\left(\frac{1}{2} x^2\right) - \int \left(\frac{1}{2} x^2\right) \frac{1}{x} dx \right]$

$= x^2 \ln x - \int x\, dx$

$= x^2 \ln x - \frac{1}{2} x^2 + C$

$= \frac{2x^2 \ln x - x^2}{2} + C$

$= \frac{x^2}{2} (\ln x^2 - 1) + C$

23. $\displaystyle\int_{1}^{2}\ln x\,dx$

$u=\ln x$	$dv=dx$
$du=\frac{1}{x}dx$	$v=x$

$\displaystyle = (\ln x)(x) - \int x\left(\tfrac{1}{x}dx\right)\Big]_{1}^{2}$

$\displaystyle = x\ln x - \int dx\Big]_{1}^{2}$

$\displaystyle = x\ln x - x\Big]_{1}^{2}$

$= (2\ln 2 - 2) - (1\ln 1 - 1)$

$= 2\ln 2 - 1 \approx 0.39$

25. $\displaystyle\int_{1}^{2}x\ln x\,dx$

$u=\ln x$	$dv=x\,dx$
$du=\frac{1}{x}dx$	$v=\frac{1}{2}x^2$

$\displaystyle = (\ln x)\left(\tfrac{1}{2}x^2\right) - \int\left(\tfrac{1}{2}x^2\right)\tfrac{1}{x}dx\Big]_{1}^{2}$

$\displaystyle = \tfrac{1}{2}x^2\ln x - \tfrac{1}{2}\int x\,dx\Big]_{1}^{2}$

$\displaystyle = \tfrac{1}{2}x\ln x - \tfrac{1}{2}\cdot\tfrac{1}{2}x^2\Big]_{1}^{2}$

$\displaystyle = \tfrac{1}{2}x^2\ln x - \tfrac{1}{4}x^2\Big]_{1}^{2}$

$\displaystyle = (2\ln 2 - 1) - \left(\tfrac{1}{2}\ln 1 - \tfrac{1}{4}\right) \approx 0.64$

27. $\displaystyle\int_{3}^{6}x(x-2)^{1/2}dx$

$u=x$	$dv=(x-2)^{1/2}dx$
$du=dx$	$v=\frac{2}{3}(x-2)^{3/2}$

$\displaystyle = x\cdot\tfrac{2}{3}(x-2)^{3/2} - \int\tfrac{2}{3}(x-2)^{3/2}dx\Big]_{3}^{6}$

$\displaystyle = \tfrac{2}{3}x(x-2)^{3/2} - \tfrac{2}{3}\cdot\tfrac{2}{5}(x-2)^{5/2}\Big]_{3}^{6}$

$\displaystyle = \tfrac{2}{3}x(\sqrt{x-2})^3 - \tfrac{4}{15}(\sqrt{x-2})^5\Big]_{3}^{6}$

$\displaystyle = \left[\tfrac{2}{3}(6)(8) - \tfrac{4}{15}(32)\right] - \left[\tfrac{2}{3}(3)(1) - \tfrac{4}{15}(1)\right] = \tfrac{326}{15}$

29. $\displaystyle\int_{0}^{1}xe^x\,dx$

$u=x$	$dv=e^x\,dx$
$du=dx$	$v=e^x$

$\displaystyle = xe^x - \int e^x\,dx\Big]_{0}^{1}$

$\displaystyle = xe^x - e^x\Big]_{0}^{1}$

$= (e^1 - e^1) - (0 - e^0) = 1$

31. $\int_0^1 \frac{x^2 e^x}{2}\,dx$

$u = \frac{1}{2}x^2$	$dv = e^x dx$
$du = x dx$	$v = e^x$

$= \frac{1}{2}x^2 e^x - \int e^x x dx \Big]_0^1$

$= \frac{1}{2}xe^x - xe^x + \int e^x dx \Big]_0^1$

$u = x$	$dv = e^x dx$
$du = dx$	$v = e^x$

$= \frac{1}{2}xe^x - xe^x + e^x \Big]_0^1$

$= \left(\frac{1}{2}e^1 - e^1 + e^1\right) - (0 - 0 + e^0)$

$= \frac{1}{2}e - 1 \approx 0.36$

33. $\ln x = 0$

$e^0 = x$

$1 = x$

$\int_1^2 \ln x\,dx$

$= (\ln x)(x) - \int x\left(\frac{1}{x}dx\right)\Big]_1^2$

$u = \ln x$	$dv = dx$
$du = \frac{1}{x}dx$	$v = x$

$= x \ln x - \int dx \Big]_1^2$

$= x \ln x - x \Big]_1^2$

$= (2 \ln 2 - 2) - (1 \ln 1 - 1) \approx 0.39$

35. $\int_0^1 xe^x dx$

$u = x$	$dv = e^x dx$
$du = dx$	$v = e^x$

$= xe^x - \int e^x dx \Big]_0^1$

$= xe^x - e^x \Big]_0^1$

$= (1e^1 - e^1) - (0 - e^0) = 1$

37. From #25 $\qquad \int_a^b x \ln x\,dx = \frac{1}{2}x^2 \ln x - \frac{1}{4}x^2 \Big]_a^b$

$\left| \int_{0.5}^1 x \ln x\,dx \right| + \int_1^2 x \ln x\,dx$

$= \left| \frac{1}{2}x^2 \ln x - \frac{1}{4}x^2 \Big]_{0.5}^1 \right| + \left[\frac{1}{2}x^2 \ln x - \frac{1}{4}x^2 \Big]_1^2 \right]$

$= \left| \left[\frac{1}{2} \ln 1 - \frac{1}{4}\right] - \left[\frac{1}{8} \ln 0.5 - \frac{1}{16}\right] \right| + [2 \ln 2 - 1] - \left[\frac{1}{2} \ln 1 - \frac{1}{4}\right] \approx 0.74$

39. $C(x) = \displaystyle\int C'(x)dx$

$u = \ln (x+1)$	$dv = dx$
$du = \dfrac{1}{x+1}dx$	$v = x$

$\quad = \displaystyle\int 2 \ln (x+1)dx$

$\quad = 2\left[x \ln (x+1) - \displaystyle\int x\dfrac{1}{x+1}dx \right]$

$\quad = 2\left[x \ln (x+1) - \displaystyle\int \dfrac{x}{x+1}dx \right]$

$\quad = 2\left[x \ln (x+1) - \displaystyle\int \left(1 - \dfrac{1}{x+1}\right)dx \right]$

$\quad = 2[x \ln (x+1) - x + \ln(x+1)] + k$

$C(0) = 2(0) \ln 1 - 2(0) + 2 \ln 1 + k$

$\quad 10 = 0 + k$

$\quad 10 = k$

$C(x) = 2x \ln (x+1) - 2x + 2 \ln(x+1) + 10$

41. $P(x) = \displaystyle\int P'(x)dx$

$u = \ln (x+1)$
$du = \dfrac{1}{x+1}dx$

$\quad = \displaystyle\int \dfrac{1000 \ln(x+1)}{x+1}dx$

$\quad = 1000 \displaystyle\int u\, du$

$\quad = 1000\dfrac{u^2}{2} + C$

$\quad = 500[\ln(x+1)]^2 + C$

$P(0) = 500[\ln(1)]^2 + C$

$\quad -100 = 0 + C$

$\quad -100 = C$

$P(x) = 500[\ln(x+1)]^2 - 100$

43a. $S(t) = \displaystyle\int S'(t)dt$

$u = \ln 2t$	$dv = dt$
$du = \dfrac{1}{2t}(2)dt$	$v = t$
$\quad = \dfrac{1}{t}dt$	

$\quad = \displaystyle\int 50 \ln 2t\, dt$

$\quad = 50\left[(\ln 2t)(t) - \displaystyle\int t\left(\dfrac{1}{t}\right)dt \right]$

$\quad = 50\left[t \ln 2t - \displaystyle\int dt \right]$

$\quad = 50t \ln 2t - 50t + C$

$$S(1) = 50(1) \ln 2 - 50 + C$$
$$450 = 50 \ln 2 - 50 + C$$
$$C \approx 465$$
$$S(t) = 50t \ln 2t - 50t + 465$$

43b. $S(6) = 50(6) \ln 12 - 50(6) + 465 \approx 910$

Section 7.2   (pages 436-438)

1. $\int \dfrac{1}{x(x+3)} dx$ Use formula 4 with $a = 3$ and $b = 1$

  $= \dfrac{-1}{3} \ln \left| \dfrac{3+x}{x} \right| + C$ or $\dfrac{1}{3} \ln \left| \dfrac{x}{3+x} \right| + C$

3. $\int \dfrac{1}{x^2 - 9} dx$ Use formula 7 with $a = 3$

  $= \dfrac{1}{2(3)} \ln \left| \dfrac{x-3}{x+3} \right| + C = \dfrac{1}{6} \ln \left| \dfrac{x-3}{x+3} \right| + C$

5. $\int \dfrac{x}{x^2 + 1} dx$ Use formula 9 with $a = 1$ and $b = 1$

  $= \dfrac{1}{2(1)} \ln \left| 1 + 1x^2 \right| + C = \dfrac{1}{2} \ln(x^2 + 1) + C$

7. $\int \sqrt{x^2 + 9} \, dx$ Use formula 16 with $a = 3$

  $= \dfrac{1}{2} \left[x \sqrt{x^2 + 9} + 9 \ln \left| x + \sqrt{x^2 + 9} \right| \right] + C$

9. $\int \dfrac{\sqrt{x^2 + 9}}{x} dx$ Use formula 19 with $a = 3$

  $= \sqrt{x^2 + 9} - 3 \ln \left| \dfrac{3 + \sqrt{x^2 + 9}}{x} \right| + C$

11. $\int \dfrac{x}{2 + 3x^2} dx$ Use formula 9 with $a = 2$ and $b = 3$

  $= \dfrac{1}{2(3)} \ln \left| 2 + 3x^2 \right| + C = \dfrac{1}{6} \ln (2 + 3x^2) + C$

13. $\int \dfrac{1}{9 - 16x^2} dx$ Use formula 12 with $a = 3$ and $b = 4$

  $= \dfrac{1}{2(3)(4)} \ln \left| \dfrac{3+4x}{3-4x} \right| + C = \dfrac{1}{24} \ln \left| \dfrac{3+4x}{3-4x} \right| + C$

15. $\displaystyle\int\frac{x}{5-2x}dx$ Use formula 1 with $a=5$ and $b=-2$

$$= \frac{x}{-2} - \frac{5}{4}\ln|5-2x| + C$$

17. $\displaystyle\int\frac{x}{(5-2x)^3}dx$ Use formula 3 with $a=5$, $b=-2$ and $n=3$

$$= \frac{1}{4}\left[\frac{-1}{(1)(5-2x)^1} + \frac{5}{2(5-2x)^2}\right] + C = \frac{1}{4}\left[\frac{-1}{(5-2x)} + \frac{5}{2(5-2x)^2}\right] + C$$

19. $\displaystyle\int x^3 \ln x\, dx$ Use formula 24 with $n=3$

$$= x^4\left[\frac{\ln x}{4} - \frac{1}{(4)^2}\right] + C = \frac{x^4 \ln x}{4} - \frac{x^4}{16} + C$$

21. $\displaystyle\int x^2 e^x dx$ Use formula 26 with $a=1$ and $n=2$

$$= \frac{x^2 e^x}{1} - \frac{2}{1}\int xe^x dx \quad\quad \text{Use formula 26 with } a=1 \text{ and } n=1$$

$$= x^2 e^x - 2\left[\frac{xe^x}{1} - \frac{1}{1}\int e^x dx\right] = x^2 e^x - 2xe^x + 2e^x + C$$

23. $\displaystyle\int\frac{1}{1+e^{2x}}dx$ Use formula 29 with $n=2$

$$= x - \frac{1}{2}\ln|1+e^{2x}| + C$$

25. $\displaystyle\int\frac{\sqrt{x+4}}{x^2}dx$ Use formula 15 with $a=4$ and $b=1$

$$= \frac{-\sqrt{4+x}}{x} + \frac{1}{2}\int\frac{1}{x\sqrt{4+x}}dx \quad\quad \text{Use formula 13 with } a=4 \text{ and } b=1$$

$$= \frac{-\sqrt{4+x}}{x} + \frac{1}{2}\cdot\frac{1}{2}\ln\left|\frac{\sqrt{4+x}-2}{\sqrt{4+x}+2}\right| + C = \frac{-\sqrt{x+4}}{x} + \frac{1}{4}\ln\left|\frac{\sqrt{x+4}-2}{\sqrt{x+4}+2}\right| + C$$

27. $\displaystyle\int_0^1 x^2 e^x dx$ (from problem #21)

$$= x^2 e^x - 2xe^x + 2e^x\Big]_0^1 = (e^1 - 2e^1 + 2e^1) - (0 - 0 + 2e^0) \approx 0.718$$

29. $\displaystyle\int_0^1\frac{x}{(x+1)^2}dx$ Use formula 2 with $a=1$ and $b=1$

$$= \frac{1}{1}\left[\frac{1}{1+x} + \ln|1+x|\right]_0^1 = \left[\frac{1}{2} + \ln 2\right] - [1 + \ln 1] \approx 0.193$$

31. $\displaystyle\int_0^4 \frac{1}{\sqrt{x^2+9}}dx$ 　　　　　Use formula 17 with $a = 3$

$= \ln\left|x + \sqrt{x^2+9}\right|\Big|_0^4 = \ln\left|4 + \sqrt{25}\right| - \ln\left|0 + \sqrt{9}\right| \approx 1.099$

33. $\displaystyle\int_{-1}^1 \frac{1}{(x^2-4)^2}dx$ 　　　　　Use formula 8 with $a = 2$

$= \dfrac{-x}{2(4)(x^2-4)} + \dfrac{1}{4(8)}\ln\left|\dfrac{2+x}{2-x}\right|\Big|_{-1}^1$

$= \left[\dfrac{-1}{8(-3)} + \dfrac{1}{32}\ln\left|\dfrac{3}{1}\right|\right] - \left[\dfrac{1}{8(-3)} + \dfrac{1}{32}\ln\left|\dfrac{1}{3}\right|\right] \approx 0.152$

35. $\displaystyle\int_1^e (\ln x)^2 dx$ 　　　　　Use formula 23 with $n = 2$

$= x(\ln x)^2 - 2\displaystyle\int \ln x\, dx\Big]_1^e$ 　　　Use formula 22

$= x(\ln x)^2 - 2(x\ln x - x)\Big|_1^e = \left[e(\ln e)^2 - 2e\ln e + 2e\right] - \left[1(\ln 1)^2 - 2\ln 1 + 2\right]$

$= (e - 2e + 2e) - 2 \approx 0.718$

37. $\displaystyle\int \frac{x^2}{1-x^6}dx$

　　　 $\boxed{\begin{array}{l} u = x^3 \\[4pt] du = 3x^2 dx \end{array}}$

$= \dfrac{1}{3}\displaystyle\int \dfrac{1}{1-(x^3)^2}(3x^2 dx)$

$= \dfrac{1}{3}\displaystyle\int \dfrac{1}{1-u^2}du$ 　　　　　Use formula 12 with $a = 1$ and $b = 1$

$= \dfrac{1}{3}\cdot\dfrac{1}{2(1)(1)}\ln\left|\dfrac{1+u}{1-u}\right| + C = \dfrac{1}{6}\ln\left|\dfrac{1+x^3}{1-x^3}\right| + C$

39. $\displaystyle\int \frac{x^3}{4+x^2}dx$

　　　 $\boxed{\begin{array}{l} u = x^2 \\[4pt] du = 2x\, dx \end{array}}$

$= \dfrac{1}{2}\displaystyle\int \dfrac{x^2}{4+x^2}(2x\, dx)$

$= \dfrac{1}{2}\displaystyle\int \dfrac{u}{4+u}du$ 　　　　　Use formula 1 with $a = 4$ and $b = 1$

$= \dfrac{1}{2}\left[\dfrac{u}{1} - \dfrac{4}{1}\ln|4+u|\right] + C = \dfrac{1}{2}x^2 - 2\ln\left|4+x^2\right| + C$

41. $\displaystyle\int \frac{x}{1+e^{x^2}}dx$

$$\boxed{\begin{array}{l} u = x^2 \\ du = 2x\,dx \end{array}}$$

$$= \frac{1}{2}\int \frac{1}{1+e^{x^2}}(2x\,dx)$$

$$= \frac{1}{2}\int \frac{1}{1+e^u}du \qquad \text{Use formula 27}$$

$$= \frac{1}{2}[u - \ln(1+e^u)] + C = \frac{1}{2}x^2 - \frac{1}{2}\ln(1+e^{x^2}) + C$$

43. $\displaystyle\int_3^5 \sqrt{x^2-9}\,dx \qquad \text{Use formula 16 with } a = 3$

$$= \frac{1}{2}\left[x\sqrt{x^2-9} - 9\ln\left|x+\sqrt{x^2-9}\right|\right]_3^5$$

$$= \frac{1}{2}\left[5\sqrt{16} - 9\ln\left|5+\sqrt{16}\right|\right] - \frac{1}{2}[3(0) - 9\ln|3+0|] \approx 5.056$$

45. $\displaystyle\int_0^1 \pi(xe^x)^2\,dx$

$$= \pi\int_0^1 x^2 e^{2x}\,dx \qquad \text{Use formula 26 with } n = 2 \text{ and } a = 2$$

$$= \pi\left[\frac{x^2 e^{2x}}{2} - \frac{2}{2}\int xe^{2x}\right]_0^1 \qquad \text{Use formula 26 with } n = 1 \text{ and } a = 2$$

$$= \pi\left[\frac{x^2 e^{2x}}{2} - \left(\frac{xe^{2x}}{2} - \frac{1}{2}\int x^0 e^{2x}dx\right)\right]_0^1 = \pi\left[\frac{x^2 e^{2x}}{2} - \frac{xe^{2x}}{2} + \frac{e^{2x}}{4}\right]_0^1$$

$$= \pi\left[\frac{e^2}{2} - \frac{e^2}{2} + \frac{e^2}{4}\right] - \pi\left[0 - 0 + \frac{e^0}{4}\right] \approx 5.018$$

47. $\displaystyle P(x) = \int_4^5\left[650 - \frac{500}{\sqrt{x^2+1}}\right]dx \qquad \text{Use formula 17 with } a = 1$

$$= 650x - 500\ln\left|x+\sqrt{x^2+1}\right|\Big|_4^5$$

$$= \left[650(5) - 500\ln\left|5+\sqrt{26}\right|\right] - \left[650(4) - 500\ln\left|4+\sqrt{17}\right|\right] \approx 541.137$$

$$\$541,137$$

49. $\displaystyle R(x) = \int R'(x)dx$

$$= \int(-xe^{-x} + e^{-x})dx$$

$$= \int e^{-x}dx - \int xe^{-x}dx \qquad \text{Use formula 26 with } a = -1 \text{ and } n = 1$$

$$= -e^{-x} - \left[\frac{xe^{-x}}{-1} - \frac{1}{-1}\int x^0 e^{-x}dx\right] = -e^{-x} + xe^{-x} - (-e^{-x}) + C$$

$$= xe^{-x} + C$$

$$R(0) = 0 + C$$
$$0 = C$$
$$R(x) = xe^{-x}$$

51.
$$N_{av} = \frac{1}{20-0}\int_0^{20}\frac{82}{4.1+15.9e^{-0.0248t}}dt$$

$$= 4.1\int_0^{20}\frac{1}{4.1+15.9e^{-0.0248t}}dt \qquad \text{Use formula 30 with } a = 4.1, \\ b = 15.9 \text{ and } n = -0.0248$$

$$= 4.1\left[\frac{t}{4.1}-\frac{1}{(4.1)(-0.0248)}\ln\left|4.1+15.9e^{-0.0248t}\right|\right]\Big|_0^{20}$$

$$= 4.1\left[\frac{20}{4.1}-\frac{1}{0.10168}\ln\left|4.1+15.9e^{-0.496}\right|\right]-4.1\left[0+\frac{1}{0.10168}\ln\left|4.1+15.9\right|\right]$$

$$\approx 4.99 \text{ billion}$$

Section 7.3 (pages 445-448)

1a. $\Delta x = \frac{4-0}{4} = 1$

$a = x_0 = 0 \qquad f(x_0) = f(0) = 0$

$x_1 = 1 \qquad f(x_1) = f(1) = 1$

$x_2 = 2 \qquad f(x_2) = f(2) = 4$

$x_3 = 3 \qquad f(x_3) = f(3) = 9$

$b = x_4 = 4 \qquad f(x_4) = f(4) = 16$

$$\int_0^4 x^2 dx \approx \frac{4-0}{2(4)}[0 + 2(1) + 2(4) + 2(9) + 16] = 22$$

1b. $\int_0^4 x^2 dx \approx \frac{4-0}{3(4)}[0 + 4(1) + 2(4) + 4(9) + 16] = \frac{64}{3}$

1c. $\int_0^4 x^2 dx = \frac{x^3}{3}\Big|_0^4 = \frac{64}{3} - 0 = \frac{64}{3}$

3a. $\Delta x = \frac{1-0}{4} = \frac{1}{4}$

$a = x_0 = 0 \qquad f(x_0) = f(0) = 1$

$x_1 = 0.25 \qquad f(x_1) = f(0.25) = 0.9375$

$x_2 = 0.5 \qquad f(x_2) = f(0.5) = 0.75$

$x_3 = 0.75 \qquad f(x_3) = f(0.75) = 0.4375$

$b = x_4 = 1 \qquad f(x_4) = f(1) = 0$

$$\int_0^1 (1-x^2)dx \approx \frac{1-0}{2(4)}[1 + 2(0.9375) + 2(0.75) + 2(0.4375) + 0] = 0.65625 = \frac{21}{32}$$

3b. $\int_0^1 (1-x^2)dx \approx \frac{1-0}{3(4)}[1 + 4(0.9375) + 2(0.75) + 4(0.4375) + 0] = \frac{8}{12} = \frac{2}{3}$

3c. $\int_0^1 (1-x^2)dx = x - \frac{1}{3}x^3 \Big|_0^1 = (1 - \frac{1}{3}) - 0 = \frac{2}{3}$

5a. $\Delta x = \frac{1-0}{8} = \frac{1}{8}$

$a = x_0 = 0 \qquad\qquad f(x_0) = f(0) = 1$

$x_1 = 0.125 \qquad f(x_1) = f(0.125) = 0.984375$

$x_2 = 0.25 \qquad f(x_2) = f(0.25) = 0.9375$

$x_3 = 0.375 \qquad f(x_3) = f(0.375) = 0.859375$

$x_4 = 0.5 \qquad f(x_4) = f(0.5) = 0.75$

$x_5 = 0.625 \qquad f(x_5) = f(0.625) = 0.609375$

$x_6 = 0.75 \qquad f(x_6) = f(0.75) = 0.4375$

$x_7 = 0.875 \qquad f(x_7) = f(0.875) = 0.234375$

$b = x_8 = 1 \qquad f(x_8) = f(1) = 0$

$$\int_0^1 (1-x^2)dx \approx \frac{1-0}{2(8)}[1 + 2(0.984375) + 2(0.9375) + 2(0.859375) + 2(0.75)$$
$$+ 2(0.609375) + 2(0.4375) + 2(0.234375) + 0]$$
$$= 0.6640625 = \frac{85}{128}$$

5b. $\int_0^1 (1-x^2)dx = \frac{1-0}{3(8)}[1 + 4(0.984375) + 2(0.9375) + 4(0.859375) + 2(0.75)$
$$+ 4(0.609375) + 2(0.4375) + 4(0.234375) + 0] = \frac{2}{3}$$

5c. $\int_0^1 (1-x^2)dx = x - \frac{1}{3}x^3 \Big|_0^1 = (1 - \frac{1}{3}) - 0 = \frac{2}{3}$

7a. $\Delta x = \frac{1-0}{6}$

$a = x_0 = 0 \qquad\qquad f(x_0) = f(0) = 1$

$x_1 = \frac{1}{6} \qquad\qquad f(x_1) = f\left(\frac{1}{6}\right) = e^{1/6}$

$x_2 = \frac{1}{3} \qquad\qquad f(x_2) = f\left(\frac{1}{3}\right) = e^{1/3}$

$x_3 = \frac{1}{2} \qquad\qquad f(x_3) = f\left(\frac{1}{2}\right) = e^{1/2}$

$x_4 = \frac{2}{3} \qquad\qquad f(x_4) = f\left(\frac{2}{3}\right) = e^{2/3}$

$x_5 = \frac{5}{6} \qquad\qquad f(x_5) = f\left(\frac{5}{6}\right) = e^{5/6}$

$b = x_6 = 1 \qquad f(x_6) = f(1) = e^1$

$$\int_0^1 e^x dx \approx \frac{1-0}{2(6)}[1 + 2e^{1/6} + 2e^{1/3} + 2e^{1/2} + 2e^{2/3} + 2e^{5/6} + e^1] \approx 1.7223$$

7b. $\int_0^1 e^x dx \approx \frac{1-0}{3(6)}[1 + 4e^{1/6} + 2e^{1/3} + 4e^{1/2} + 2e^{2/3} + 4e^{5/6} + e^1] \approx 1.7182889$

7c. $\int_0^1 e^x dx = e^x \Big]_0^1 = e^1 - e^0 \approx 2.7182818 - 1 = 1.7182818$

9a.
$$\Delta x = \frac{4-1}{6} = \frac{1}{2}$$

$a = x_0 = 1$	$f(x_0) = f(1) = 1$
$x_1 = 1.5$	$f(x_1) = f(1.5) = \frac{2}{3}$
$x_2 = 2$	$f(x_2) = f(2) = \frac{1}{2}$
$x_3 = 2.5$	$f(x_3) = f(2.5) = \frac{2}{5}$
$x_4 = 3$	$f(x_4) = f(3) = \frac{1}{3}$
$x_5 = 3.5$	$f(x_5) = f(3.5) = \frac{2}{7}$
$b = x_6 = 4$	$f(x_6) = f(4) = \frac{1}{4}$

$$\int_1^4 \frac{1}{x} dx \approx \frac{4-1}{2(6)}\left[1 + 2\left(\frac{2}{3}\right) + 2\left(\frac{1}{2}\right) + 2\left(\frac{2}{5}\right) + 2\left(\frac{1}{3}\right) + 2\left(\frac{2}{7}\right) + \frac{1}{4}\right] \approx 1.4054$$

9b. $\int_1^4 \frac{1}{x} dx \approx \frac{4-1}{3(6)}\left[1 + 4\left(\frac{2}{3}\right) + 2\left(\frac{1}{2}\right) + 4\left(\frac{2}{5}\right) + 2\left(\frac{1}{3}\right) + 4\left(\frac{2}{7}\right) + \frac{1}{4}\right] \approx 1.3877$

9c. $\int_1^4 \frac{1}{x} dx = \ln|x| \Big]_1^4 = \ln 4 - \ln 1 \approx 1.3863$

11a.
$$\Delta x = \frac{4-0}{4} = 1$$

$a = x_0 = 0$	$f(x_0) = f(0) = 0$
$x_1 = 1$	$f(x_1) = f(1) = 1$
$x_2 = 2$	$f(x_2) = f(2) = \sqrt{2}$
$x_3 = 3$	$f(x_3) = f(3) = \sqrt{3}$
$b = x_4 = 4$	$f(x_4) = f(4) = 2$

$$\int_0^4 \sqrt{x} dx \approx \frac{4-0}{2(4)}\left[0 + 2(1) + 2\sqrt{2} + 2\sqrt{3} + 2\right] \approx 5.1463$$

11b. $\int_0^4 \sqrt{x} dx \approx \frac{4-0}{3(4)}\left[0 + 4(1) + 2\sqrt{2} + 4\sqrt{3} + 2\right] \approx 5.2522$

11c. $\int_0^4 \sqrt{x} dx = \int_0^4 x^{1/2} dx = \frac{2}{3}x^{3/2} \Big]_0^4 = \frac{2}{3}\left(\sqrt{4}\right)^3 - 0 \approx 5.3333$

13a.

$$\Delta x = \frac{1-0}{4} = \frac{1}{4}$$

$$a = x_0 = 0 \qquad\qquad f(x_0) = f(0) = 1$$

$$x_1 = \frac{1}{4} \qquad\qquad f(x_1) = f\left(\frac{1}{4}\right) = e^{1/16}$$

$$x_2 = \frac{1}{2} \qquad\qquad f(x_2) = f\left(\frac{1}{2}\right) = e^{1/4}$$

$$x_3 = \frac{3}{4} \qquad\qquad f(x_3) = f\left(\frac{3}{4}\right) = e^{9/16}$$

$$b = x_4 = 1 \qquad\qquad f(x_4) = f(1) = e^1$$

$$\int_0^1 e^{x^2} dx \approx \frac{1-0}{2(4)}[1 + 2e^{1/16} + 2e^{1/4} + 2e^{9/16} + e^1] \approx 1.4907$$

13b. $\quad \int_0^1 e^{x^2} dx \approx \frac{1-0}{3(4)}[1 + 4e^{1/16} + 2e^{1/4} + 4e^{9/16} + e^1] \approx 1.4637$

15a.

$$\Delta x = \frac{2-1}{4} = \frac{1}{4}$$

$$a = x_0 = 1 \qquad\qquad f(x_0) = f(1) = 0$$

$$x_1 = 1.25 \qquad\qquad f(x_1) = f(1.25) = \sqrt{\frac{61}{64}}$$

$$x_2 = 1.5 \qquad\qquad f(x_2) = f(1.5) = \sqrt{\frac{19}{8}}$$

$$x_3 = 1.75 \qquad\qquad f(x_3) = f(1.75) = \sqrt{\frac{279}{64}}$$

$$b = x_4 = 2 \qquad\qquad f(x_4) = f(2) = \sqrt{7}$$

$$\int_1^2 \sqrt{x^3 - 1}\, dx = \frac{2-1}{2(4)}\left[0 + 2\sqrt{\frac{61}{64}} + 2\sqrt{\frac{19}{8}} + 2\sqrt{\frac{279}{64}} + \sqrt{7}\right] \approx 1.4820$$

15b. $\quad \int_1^2 \sqrt{x^3 - 1}\, dx = \frac{2-1}{3(4)}\left[0 + 4\sqrt{\frac{61}{64}} + 2\sqrt{\frac{19}{8}} + 4\sqrt{\frac{279}{64}} + \sqrt{7}\right] \approx 1.4987$

17a.

$$\Delta x = \frac{1-0}{4} = \frac{1}{4}$$

$$a = x_0 = 0 \qquad\qquad f(x_0) = f(0) = 1$$

$$x_1 = 0.25 \qquad\qquad f(x_1) = f(0.25) = \frac{1}{\sqrt{1.015625}}$$

$$x_2 = 0.5 \qquad\qquad f(x_2) = f(0.5) = \frac{1}{\sqrt{1.125}}$$

$$x_3 = 0.75 \qquad\qquad f(x_3) = f(0.75) = \frac{1}{\sqrt{1.421875}}$$

$$b = x_4 = 1 \qquad\qquad f(x_4) = f(1) = \frac{1}{\sqrt{2}}$$

$$\int_0^1 \frac{1}{\sqrt{x^3 + 1}}\, dx = \frac{1-0}{2(4)}\left[1 + \frac{2}{\sqrt{1.015625}} + \frac{2}{\sqrt{1.125}} + \frac{2}{\sqrt{1.421875}} + \frac{1}{\sqrt{2}}\right] \approx 0.9068$$

17b. $\displaystyle\int_0^1 \frac{1}{\sqrt{x^3+1}}dx = \frac{1-0}{3(4)}\left[1 + \frac{4}{\sqrt{1.015625}} + \frac{2}{\sqrt{1.125}} + \frac{4}{\sqrt{1.421875}} + \frac{1}{\sqrt{2}}\right] \approx 0.9097$

19. $\displaystyle\frac{6-0}{2(6)}[f(0) + 2f(1) + 2f(2) + 2f(3) + 2f(4) + 2f(5) + f(6)]$

$= \frac{1}{2}[2 + 2(4) + 2(6) + 2(5) + 2(2) + 2(3) + 2] = 22$

21. $\displaystyle\frac{10-1}{2(9)}[P(1) + 2P(2) + 2P(3) + 2P(4) + 2P(5) + 2P(6) + 2P(7) + 2P(8)$

$+ 2P(9) + P(10)]$

$= \frac{1}{2}[12.4 + 2(15.1) + 2(15.4) + 2(17.0) + 2(20.5) + 2(20.5) + 2(21.0) + 2(24.0)$

$+ 2(24.2) + 25] = 176.4 \qquad \$176{,}400$

23. $\displaystyle\frac{6-0}{2(6)}[H(0) + 2H(1) + 2H(2) + 2H(3) + 2H(4) + 2H(5) + H(6)]$

$= \frac{1}{2}[2 + 2(0.2) + 2(0.1) + 2(0.1) + 2(0.2) + 2(0.5) + 0.3] = 1.25 \qquad 125 \text{ homes}$

25. If $x = 5$, $p = \sqrt{25 + 5 + 1} = \sqrt{31}$

producer surplus $= \displaystyle\int_0^5 \left[\sqrt{31} - \sqrt{x^2 + x + 1}\right]dx$

$\Delta x = \dfrac{5-0}{4} = \dfrac{5}{4}$

$\begin{array}{ll} a = x_0 = 0 & f(x_0) = f(0) \approx 4.568 \\ x_1 = 1.25 & f(x_1) = f(1.25) \approx 3.615 \\ x_2 = 2.5 & f(x_2) = f(2.5) \approx 2.445 \\ x_3 = 3.75 & f(x_3) = f(3.75) \approx 1.230 \\ b = x_4 = 5 & f(x_4) = f(5) = 0 \end{array}$

$\displaystyle\int_0^5 \left[\sqrt{31} - \sqrt{x^2 + x + 1}\right]dx \approx \frac{5-0}{3(4)}[4.568 + 4(3.615) + 2(2.445) + 4(1.230) + 0]$

≈ 12.02

27. $\Delta t = \dfrac{6-0}{6} = 1$

$\begin{array}{ll} a = t_0 = 0 & B(t_0) = B(0) = 1 \\ t_1 = 1 & B(t_1) = B(1) = e^{-0.2} \\ t_2 = 2 & B(t_2) = B(2) = e^{-0.8} \\ t_3 = 3 & B(t_3) = B(3) = e^{-1.8} \\ t_4 = 4 & B(t_4) = B(4) = e^{-3.2} \\ t_5 = 5 & B(t_5) = B(5) = e^{-5} \\ b = t_6 = 6 & B(t_6) = B(6) = e^{-7.2} \end{array}$

$$\int_0^6 e^{-0.2t^2} dt \approx \frac{6-0}{3(6)}\left[1 + 4e^{-0.2} + 2e^{-0.8} + 4e^{-1.8} + 2e^{-3.2} + 4e^{-5} + e^{-7.2}\right]$$
$$\approx 2 \text{ babies}$$

29. $\dfrac{500-0}{3(10)}[0.39 + 4(0.27) + 2(0.25) + 4(0.23) + 2(0.22) + 4(0.21) + 2(0.20) + 4(0.19)$
$+ 2(0.18) + 4(0.17) + 0.16] \approx 108.83$

31. $f(x) = x^{1/2}$

$f'(x) = \frac{1}{2}x^{-1/2}$

$f''(x) = \frac{-1}{4}x^{-3/2} = \frac{-1}{4\sqrt{x^3}}$

$\left|f''(x)\right| = \left|\dfrac{-1}{4\sqrt{x^3}}\right|$

$M = \left|f''(1)\right| = \left|\dfrac{-1}{4(1)}\right| = \frac{1}{4}$

$\dfrac{(2-1)^3\left(\frac{1}{4}\right)}{12n^2} \le 0.001$

$\dfrac{1}{n^2} \le \dfrac{0.012}{0.25}$

$n^2 \ge \dfrac{0.25}{0.012}$

$n \ge 4.56$

Use $n = 5$.

33. $f(x) = x^{-2}$

$f'(x) = -2x^{-3}$

$f''(x) = 6x^{-4}$

$f'''(x) = -24x^{-5}$

$f^{(4)}(x) = 120x^{-6}$

$\left|f^{(4)}(x)\right| = \left|\dfrac{120}{x^6}\right|$

$M = \left|f^{(4)}(1)\right| = 120$

$\dfrac{(2-1)^5(120)}{180n^4} \le 0.002$

$\dfrac{1}{n^4} \le 0.003$

$n^4 \ge \dfrac{1000}{3}$

$n \ge 4.27$

Since n must be even, use $n = 6$.

Section 7.3A   (pages 454-455)

1. area ≈ 1.098660599

$\displaystyle\int_1^3 \frac{1}{x}dx = \ln|x| \Big|_1^3 = \ln 3 - \ln 1 \approx 1.098612289$

3. area ≈ 1.098612373

$\displaystyle\int_1^3 \frac{1}{x}dx = \ln|x| \Big|_1^3 = \ln 3 - \ln 1 \approx 1.098612289$

5. area ≈ 0.7668301685

$\displaystyle\int_1^4 x^{-2}dx = -x^{-1} \Big|_1^4 = \frac{-1}{4} - (-1) = 0.75$

7. area ≈ 5.252210118

$$\int_0^4 x^{1/2}\,dx = \frac{x^{3/2}}{\frac{3}{2}}\Big]_0^4 = \frac{2}{3}\sqrt{x^3}\Big]_0^4 = \frac{2}{3}(8) - 0 = \frac{16}{3}$$

9. area ≈ 1457.76818 **11.** area ≈ 11.82056658

13. area ≈ 2.541116664 **15.** area ≈ 15.78666667

17. $(x_0,y_0) = (0,700)$ $(x_1,y_1) = (1,1428)$ $(x_2,y_2) = (2,700)$

$700 = A(0)^2 + B(0) + C$ $C = 700$

$1428 = A(1)^2 + B(1) + C$ $728 = A + B$

$700 = A(2)^2 + B(2) + C$ $0 = 4A + 2B$

Solving the equations simultaneously, $A = -728$, $B = 1456$

$y = -728x^2 + 1456x + 700$

19a. $x_1 = x_0 + h$ $x_2 = x_1 + h$

$x_1 - h = x_0$ so $\displaystyle\int_{x_0}^{x_2}(Ax^2 + Bx + C)\,dx = \int_{x_1-h}^{x_1+h}(Ax^2 + Bx + C)\,dx$

19b. $\displaystyle\int_{x_1-h}^{x_1+h}(Ax^2 + Bx + C)\,dx = \frac{Ax^3}{3} + \frac{Bx^2}{2} + Cx\Big]_{x_1-h}^{x_1+h}$

$$= \left[\frac{A(x_1+h)^3}{3} + \frac{B(x_1+h)^2}{2} + C(x_1+h)\right] - \left[\frac{A(x_1-h)^3}{3} + \frac{B(x_1-h)^2}{2} + C(x_1-h)\right]$$

$$= \frac{A(x_1^3 + 3x_1^2 h + 3x_1 h^2 + h^3)}{3} + \frac{B(x_1^2 + 2x_1 h + h^2)}{2} + C(x_1 + h)$$

$$- \frac{A(x_1^3 - 3x_1^2 h + 3x_1 h^2 - h^3)}{3} - \frac{B(x_1^2 - 2x_1 h + h^2)}{2} - C(x_1 - h)$$

$$= \frac{A(6x_1^2 h + 2h^3)}{3} + \frac{B(4x_1 h)}{2} + C(2h) = \frac{A}{3}(6hx_1^2 + 2h^3) + 2h(Bx_1 + C)$$

19c. $\dfrac{A}{3}(6hx_1^2 + 2h^3) + 2h(Bx_1 + C) = \dfrac{A}{3}(6hx_1^2 + 2h^3) + 2h(y_1 - Ax_1^2)$

$$= 2Ahx_1^2 + \frac{2}{3}Ah^3 + 2hy_1 - 2Ahx_1^2) = 2h\left(\frac{Ah^2}{3} + y_1\right)$$

$$= 2h\left[\frac{h^2}{3}\cdot\frac{y_0 - 2y_1 + y_2}{2h^2} + y_1\right] = \frac{h}{3}(y_0 - 2y_1 + y_2) + 2hy_1$$

$$= \frac{h}{3}(y_0 - 2y_1 + y_2) + \frac{6h}{3}y_1 = \frac{h}{3}(y_0 - 2y_1 + y_2 + 6y_1)$$

$$= \frac{h}{3}(y_0 + 4y_1 + y^2)$$

21. $\left[\frac{-728}{3}x^3 + 728x^2 + 700x\right]_0^2 + \left[\frac{112}{3}x^3 - 280x^2 + 1372x\right]_2^4$

$+ \left[\frac{-104}{3}x^3 + 356x^2 - 260x\right]_4^6 + \left[\frac{64}{3}x^3 - 340x^2 + 2044x\right]_6^8$

$= 2370\frac{2}{3} + \left(3397\frac{1}{3} - 1922\frac{2}{3}\right) + \left(3768 - 2437\frac{1}{3}\right) + \left(5514\frac{2}{3} - 4632\right) = 6058\frac{2}{3}$

This is the same value found by using the calculator program in this section.

Section 7.4 (pages 461-463)

1. $\displaystyle\int_1^\infty 4x^{-2}dx = \lim_{b\to\infty}\left[\int_1^b 4x^{-2}dx\right]$

$= \lim_{b\to\infty}\frac{4x^{-1}}{-1}\Big|_1^b$

$= \lim_{b\to\infty}\frac{-4}{x}\Big|_1^b$

$= \lim_{b\to\infty}\left[\frac{-4}{b} - \frac{-4}{1}\right]$

$= 0 + 4 = 4$

3. $\displaystyle\int_8^\infty 6x^{-1/3}dx = \lim_{b\to\infty}\left[\int_8^b 6x^{-1/3}dx\right]$

$= \lim_{b\to\infty}6\frac{x^{2/3}}{\frac{2}{3}}\Big|_8^b$

$= \lim_{b\to\infty}9\sqrt[3]{x^2}\Big|_8^b$

$= \lim_{b\to\infty}\left[9\sqrt[3]{b^2} - 9(4)\right]$

diverges

5. $\displaystyle\int_2^\infty 4(x-1)^{-1/2}dx = \lim_{b\to\infty}\left[\int_2^b 4(x-1)^{-1/2}dx\right]$

$= \lim_{b\to\infty}\left[4\frac{(x-1)^{1/2}}{\frac{1}{2}}\right]_2^b$

$= \lim_{b\to\infty}\left[8\sqrt{x-1}\right]_2^b$

$= \lim_{b\to\infty}\left[8\sqrt{b-1} - 8(1)\right]$

diverges

7. $\displaystyle\int_3^\infty (x-2)^{-2}dx = \lim_{b\to\infty}\left[\int_3^b (x-2)^{-2}dx\right]$

$= \lim_{b\to\infty}\left[\frac{(x-2)^{-1}}{-1}\right]_3^b$

$= \lim_{b\to\infty}\left[\frac{-1}{x-2}\right]_3^b$

$= \lim_{b\to\infty}\left[\frac{-1}{b-2} - (-1)\right]$

$= 0 + 1 = 1$

9. $\displaystyle\int (4-x)^{-1/2}dx$ $\boxed{\begin{array}{l} u = 4-x \\ du = -dx \end{array}}$

$\displaystyle = -\int (4-x)^{-1/2}(-dx)$

$\displaystyle = -\int u^{-1/2}du$

$\displaystyle = \frac{-u^{1/2}}{\frac{1}{2}}+C$

$\displaystyle = -2\sqrt{4-x}+C$

$\displaystyle\int_{-\infty}^{3}(4-x)^{-1/2}dx = \lim_{a\to-\infty}\left[\int_{a}^{3}(4-x)^{-1/2}dx\right]$

$\displaystyle\qquad\qquad = \lim_{a\to-\infty}\left[-2\sqrt{4-x}\,\right]_{a}^{3}$

$\displaystyle\qquad\qquad = \lim_{a\to-\infty}\left[-2-(-2\sqrt{4-a})\right]$

diverges

11. $\displaystyle\int 2x(x^2+1)^{-1/2}dx$ $\boxed{\begin{array}{l} u = x^2+1 \\ du = 2xdx \end{array}}$

$\displaystyle = \int u^{-1/2}du$

$\displaystyle = \frac{u^{1/2}}{\frac{1}{2}}+C$

$\displaystyle = 2\sqrt{x^2+1}+C$

$\displaystyle\int_{0}^{\infty}2x(x^2+1)^{-1/2}dx = \lim_{b\to\infty}\left[\int_{0}^{b}2x(x^2+1)^{-1/2}dx\right]$

$\displaystyle\qquad\qquad = \lim_{b\to\infty}\left[2\sqrt{x^2+1}\,\right]_{0}^{b}$

$\displaystyle\qquad\qquad = \lim_{b\to\infty}\left[2\sqrt{b^2+1}-2\right]$

diverges

13. $\displaystyle\int_{-\infty}^{-2}5(x+1)^{-2}dx = \lim_{a\to-\infty}\left[\int_{a}^{-2}5(x+1)^{-2}dx\right]$

$\displaystyle\qquad\qquad = \lim_{a\to-\infty}\left[5\frac{(x+1)^{-1}}{-1}\right]_{a}^{-2}$

$\displaystyle\qquad\qquad = \lim_{a\to-\infty}\left[\frac{-5}{x+1}\right]_{a}^{-2}$

$\displaystyle\qquad\qquad = \lim_{a\to-\infty}\left[\frac{-5}{-1}-\frac{-5}{a+1}\right]$

$\displaystyle\qquad\qquad = 5-0 = 5$

15. $\displaystyle\int (2x-1)^{-3} dx$

$\boxed{\begin{array}{l} u = 2x - 1 \\ du = 2dx \end{array}}$

$= \dfrac{1}{2}\displaystyle\int (2x-1)^{-3}(2dx)$

$= \dfrac{1}{2}\displaystyle\int u^{-3} du$

$= \dfrac{1}{2}\dfrac{u^{-2}}{-2} + C$

$= \dfrac{-1}{4(2x-1)} + C$

$\displaystyle\int_1^\infty (2x-1)^{-3} dx = \lim_{b\to\infty}\left[\int_1^b (2x-1)^{-3} dx\right]$

$\qquad\qquad = \lim_{b\to\infty}\left[\dfrac{-1}{4(2x-1)}\right]_1^b$

$\qquad\qquad = \lim_{b\to\infty}\left[\dfrac{-1}{4(2b-1)} - \dfrac{-1}{4}\right]$

$\qquad\qquad = 0 + \dfrac{1}{4} = \dfrac{1}{4}$

17. $\displaystyle\int (4x-3)^{-1/2} dx$

$\boxed{\begin{array}{l} u = 4x - 3 \\ du = 4dx \end{array}}$

$= \dfrac{1}{4}\displaystyle\int (4x-3)^{-1/2}(4dx)$

$= \dfrac{1}{4}\displaystyle\int u^{-1/2} du$

$= \dfrac{1}{4}\dfrac{u^{1/2}}{\frac{1}{2}} + C$

$= \dfrac{1}{2}\sqrt{4x-3} + C$

$\displaystyle\int_1^\infty (4x-3)^{-1/2} dx = \lim_{b\to\infty}\left[\int_1^b (4x-3)^{-1/2} dx\right]$

$\qquad\qquad = \lim_{b\to\infty}\left[\dfrac{1}{2}\sqrt{4x-3}\right]_1^b$

$\qquad\qquad = \lim_{b\to\infty}\left[\dfrac{1}{2}\sqrt{4b-3} - \dfrac{1}{2}\right]$

diverges

19.

$$\int (x^2+1)^{-1/3} 8x\,dx \qquad \boxed{\begin{array}{l} u = x^2+1 \\ du = 2x\,dx \end{array}}$$

$$= 4\int (x^2+1)^{-1/3}(2x\,dx)$$

$$= 4\int u^{-1/3}\,du$$

$$= 4\frac{u^{2/3}}{\frac{2}{3}} + C$$

$$= 6\sqrt[3]{(x^2+1)^2} + C$$

$$\int_{-\infty}^{\infty} (x^2+1)^{-1/3} 8x\,dx = \int_{-\infty}^{0} (x^2+1)^{-1/3} 8x\,dx + \int_{0}^{\infty} (x^2+1)^{-1/3} 8x\,dx$$

$$= \lim_{a\to -\infty}\left[\int_{a}^{0}(x^2+1)^{-1/3} 8x\,dx\right] + \lim_{b\to \infty}\left[\int_{0}^{b}(x^2+1)^{-1/3} 8x\,dx\right]$$

$$= \lim_{a\to -\infty}\left[6\sqrt[3]{(x^2+1)^2}\Big|_{a}^{0}\right] + \lim_{b\to \infty}\left[6\sqrt[3]{(x^2+1)^2}\Big|_{0}^{b}\right]$$

$$= \lim_{a\to -\infty}\left[6 - 6\sqrt[3]{(a^2+1)^2}\right] + \lim_{b\to \infty}\left[6\sqrt[3]{(b^2+1)^2} - 6\right]$$

diverges

21.

$$\int_{-\infty}^{0} e^{3x}\,dx = \lim_{a\to -\infty}\left[\int_{a}^{0} e^{3x}\,dx\right]$$

$$= \lim_{a\to -\infty}\left[\frac{e^{3x}}{3}\Big|_{a}^{0}\right]$$

$$= \lim_{a\to -\infty}\left[\frac{1}{3} - \frac{e^{3a}}{3}\right]$$

$$= \frac{1}{3} - 0 = \frac{1}{3}$$

23.

$$\int 2xe^{-x^2}\,dx \qquad \boxed{\begin{array}{l} u = -x^2 \\ du = -2x\,dx \end{array}}$$

$$= -\int e^{-x^2}(-2x\,dx)$$

$$= -\int e^{u}\,du$$

$$= -e^{u} + C$$

$$= -e^{-x^2} + C$$

$$\int_{0}^{\infty} 2xe^{-x^2}\,dx = \lim_{b\to \infty}\left[\int_{0}^{b} 2xe^{-x^2}\,dx\right]$$

$$= \lim_{b\to \infty}\left[-e^{-x^2}\Big|_{0}^{b}\right]$$

$$= \lim_{b\to \infty}\left[-e^{-b^2} - (-1)\right]$$

$$= 0 + 1 = 1$$

25.

$$\int_1^\infty \frac{1}{x+3}dx = \lim_{b\to\infty}\left[\int_1^b \frac{1}{1+3}dx\right]$$

$$= \lim_{b\to\infty}\left[\ln|x+3|\Big|_1^b\right]$$

$$= \lim_{b\to\infty}\left[\ln|b+3|-\ln 4\right]$$

diverges

27.

$$\int \frac{1}{x^2+1}(4x)dx \qquad \boxed{\begin{array}{l} u = x^2+1 \\ du = 2xdx \end{array}}$$

$$= 2\int \frac{1}{x^2+1}(2xdx)$$

$$= 2\int \frac{1}{u}du$$

$$= 2\ln|u|+C$$

$$= 2\ln(x^2+1)+C$$

$$\int_0^\infty \frac{1}{x^2+1}(4x)dx = \lim_{b\to\infty}\left[\int_0^b \frac{1}{x^2+1}(4x)dx\right]$$

$$= \lim_{b\to\infty}\left[2\ln(x^2+1)\Big|_0^b\right]$$

$$= \lim_{b\to\infty}\left[2\ln(b^2+1)-0\right]$$

diverges

29.

$y = \frac{2}{x^2}$

$$\int_1^\infty \frac{2}{x^2}dx = \lim_{b\to\infty}\left[\int_1^b 2x^{-2}dx\right]$$

$$= \lim_{b\to\infty}\left[2\frac{x^{-1}}{-1}\Big|_1^b\right]$$

$$= \lim_{b\to\infty}\left[\frac{-2}{b}-\frac{-2}{1}\right]$$

$$= 0+2 = 2$$

31.

$y = e^{-x}$

$y = -e^{-x}$

$$\int_1^\infty [e^{-x}-(-e^{-x})]dx = \lim_{b\to\infty}\left[\int_1^b (e^{-x}+e^{-x})dx\right]$$

$$= \lim_{b\to\infty}\left[\int_1^b 2e^{-x}dx\right]$$

$$= \lim_{b\to\infty}\left[2\frac{e^{-x}}{-1}\Big|_1^b\right]$$

$$= \lim_{b\to\infty}\left[-2e^{-b}-(-2e^{-1})\right]$$

$$= 0+2e^{-1} \approx 0.736$$

33. $\left| \int_{-\infty}^{0} \dfrac{2x}{(x^2+1)^4} dx \right| + \int_{0}^{\infty} \dfrac{2x}{(x^2+1)^4} dx = \left| \dfrac{-1}{3} \right| + \dfrac{1}{3} = \dfrac{1}{3} + \dfrac{1}{3} = \dfrac{2}{3}$

35.
$$\int_{0}^{\infty} 15000e^{-0.08x} dx = \lim_{b \to \infty} \left[\int_{0}^{b} 15000e^{-0.08x} dx \right]$$
$$= \lim_{b \to \infty} \left[15000 \dfrac{e^{-0.08x}}{-0.08} \Big|_{0}^{b} \right]$$
$$= \lim_{b \to \infty} \left[-187,500e^{-0.08x} \Big|_{0}^{b} \right]$$
$$= \lim_{b \to \infty} \left[-187,500e^{-0.08b} + 187,500 \right]$$
$$= 0 + 187,500 = \$187,500$$

37.
$$\int_{0}^{\infty} \dfrac{1}{t+0.01} dt = \lim_{b \to \infty} \left[\int_{0}^{b} \dfrac{1}{t+0.01} dt \right]$$
$$= \lim_{b \to \infty} \left[\ln |t+0.01| \Big|_{0}^{b} \right]$$
$$= \lim_{b \to \infty} \left[\ln |b+0.01| - \ln(0.01) \right]$$
diverges (Pollutants increase without bound.)

Chapter 7 Review   (pages 464-465)

1. $\displaystyle\int \ln 4x \, dx$

$u = \ln 4x$	$dv = dx$
$du = \dfrac{1}{4x}(4)dx$	$v = x$
$= \dfrac{1}{x}dx$	

$= (\ln 4x)(x) - \displaystyle\int x\left(\dfrac{1}{x}dx\right)$

$= x \ln 4x - \displaystyle\int dx$

$= x \ln 4x - x + C$

2. $\displaystyle\int \ln x^{1/2} \, dx$

$u = \ln x$	$dv = dx$
$du = \dfrac{1}{x}dx$	$v = x$

$= \dfrac{1}{2} \displaystyle\int \ln x \, dx$

$= \dfrac{1}{2} \left[(\ln x)(x) - \displaystyle\int x\left(\dfrac{1}{x}dx\right) \right]$

$= \dfrac{1}{2} \left[x \ln x - \displaystyle\int dx \right]$

$= \dfrac{1}{2}x \ln x - \dfrac{1}{2}x + C$

$= \dfrac{x}{2}(\ln x - 1) + C$

3. $\displaystyle\int \frac{3x}{(1+2x)^2}dx$ Use formula 2 with $a=1$ and $b=2$

$\displaystyle = 3\int \frac{x}{(1+2x)^2}dx$

$\displaystyle = 3\left(\frac{1}{4}\right)\left[\frac{1}{1+2x}+\ln|1+2x|\right]+C$

$\displaystyle = \frac{3}{4}\left[\frac{1}{1+2x}+\ln|1+2x|\right]+C$

4. $\displaystyle\int \frac{1}{9x^2-16}dx$

$\displaystyle = \int \frac{1}{(3x)^2-4^2}dx$

$u=3x$
$du=3dx$

$\displaystyle = \frac{1}{3}\int \frac{1}{(3x)^2-4^2}(3dx)$

$\displaystyle = \frac{1}{3}\int \frac{1}{u^2-4^2}du$ Use formula 7 with $a=4$

$\displaystyle = \frac{1}{3}\left[\frac{1}{2(4)}\ln\left|\frac{u-4}{u+4}\right|\right]+C$

$\displaystyle = \frac{1}{24}\ln\left|\frac{3x-4}{3x+4}\right|+C$

5. $\displaystyle\int xe^{4x}\,dx$

$u=x$	$dv=e^{4x}\,dx$
$du=dx$	$v=\dfrac{e^{4x}}{4}$

$\displaystyle = x\left(\frac{e^{4x}}{4}\right)-\int \frac{e^{4x}}{4}dx$

$\displaystyle = \frac{xe^{4x}}{4}-\frac{1}{4}\frac{e^{4x}}{4}+C$

$\displaystyle = \frac{e^{4x}}{4}-\frac{e^{4x}}{16}+C$

$\displaystyle = \frac{e^{4x}}{16}(4x-1)+C$

6. $\displaystyle\int 2xe^{-x}\,dx$

$u=2x$	$dv=e^{-x}\,dx$
$du=2dx$	$v=-e^{-x}$

$\displaystyle = 2x(-e^{-x})-\int(-e^{-x})(2dx)$

$\displaystyle = -2xe^{-x}+2\int e^{-x}dx$

$\displaystyle = -2xe^{-x}+2(-e^{-x})+C$

$\displaystyle = -2e^{-x}(x+1)+C$

7. $\displaystyle\int \frac{1}{1+e^{2x}}dx$ Use formula 29 with $n=2$

$\displaystyle = x-\frac{1}{2}\ln|1+e^{2x}|+C$

8. $\int \frac{\sqrt{x^2+9}}{x}dx$ Use formula 19 with $a = 3$

$= \sqrt{x^2+9} - 3\ln\left|\frac{3+\sqrt{x^2+9}}{x}\right| + C$

9. $\int_0^1 xe^{2x}dx$

$u = x$	$dv = e^{2x}\,dx$
$du = dx$	$v = \frac{e^{2x}}{2}$

$= x\left(\frac{e^{2x}}{2}\right) - \int \frac{e^{2x}}{2}dx\Big]_0^1$

$= \frac{xe^{2x}}{2} - \frac{1}{2}\frac{e^{2x}}{2}\Big]_0^1$

$= \frac{xe^{2x}}{2} - \frac{e^{2x}}{4}\Big]_0^1$

$= \left(\frac{e^2}{2} - \frac{e^2}{4}\right) - \left(0 - \frac{1}{4}\right) \approx 2.098$

10. $\int_1^3 x\ln x\,dx$

$u = \ln x$	$dv = x\,dx$
$du = \frac{1}{x}dx$	$v = \frac{x^2}{2}$

$= (\ln x)\left(\frac{x^2}{2}\right) - \int \frac{x^2}{2}\cdot\frac{1}{x}dx\Big]_1^3$

$= \frac{x^2\ln x}{2} - \frac{1}{2}\int x\,dx\Big]_1^3$

$= \frac{x^2\ln x}{2} - \frac{1}{2}\cdot\frac{x^2}{2}\Big]_1^3$

$= \frac{x^2\ln x}{2} - \frac{x^2}{4}\Big]_1^3$

$= \left(\frac{9\ln 3}{2} - \frac{9}{4}\right) - \left(0 - \frac{1}{4}\right) \approx 2.944$

11. $\int_0^1 \frac{1}{9-x^2}dx$ Use formula 12 with $a = 3$ and $b = 1$

$= \frac{1}{2(3)(1)}\ln\left|\frac{3+x}{3-x}\right|\Big]_0^1$

$= \frac{1}{6}\left[\ln\frac{4}{2} - \ln\frac{3}{3}\right] \approx 0.116$

12. $\int_0^3 x\sqrt{x+1}dx$

$u = x$	$dv = (x+1)^{1/2}\,dx$
$du = dx$	$v = \frac{2}{3}(x+1)^{3/2}$

$= x\cdot\frac{2}{3}(x+1)^{3/2} - \int \frac{2}{3}(x+1)^{3/2}dx\Big]_0^3$

$= \frac{2}{3}x(x+1)^{3/2} - \frac{2}{3}\cdot\frac{2}{5}(x+1)^{5/2}\Big]_0^3$

$= \left[\frac{2}{3}(3)(8) - \frac{4}{15}(32)\right] - \left[0 - \frac{4}{15}\right] \approx 7.733$

13a. $\Delta x = \frac{3-1}{8} = \frac{1}{4}$

$a = x_0 = 1$ $\qquad\qquad f(x_0) = f(1) = 1$

$\qquad x_1 = 1.25$ $\qquad f(x_1) = f(1.25) = \frac{4}{5}$

$\qquad x_2 = 1.5$ $\qquad f(x_2) = f(1.5) = \frac{2}{3}$

$\qquad x_3 = 1.75$ $\qquad f(x_3) = f(1.75) = \frac{4}{7}$

$\qquad x_4 = 2$ $\qquad\quad f(x_4) = f(2) = \frac{1}{2}$

$\qquad x_5 = 2.25$ $\qquad f(x_5) = f(2.25) = \frac{4}{9}$

$\qquad x_6 = 2.5$ $\qquad f(x_6) = f(2.5) = \frac{2}{5}$

$\qquad x_7 = 2.75$ $\qquad f(x_7) = f(2.75) = \frac{4}{11}$

$b = x_8 = 3$ $\qquad\quad f(x_8) = f(3) = \frac{1}{3}$

$\int_1^3 \frac{1}{x} dx \approx \frac{3-1}{2(8)}\left[1 + 2\left(\frac{4}{5}\right) + 2\left(\frac{2}{3}\right) + 2\left(\frac{4}{7}\right) + 2\left(\frac{1}{2}\right) + 2\left(\frac{4}{9}\right) + 2\left(\frac{2}{5}\right) + 2\left(\frac{4}{11}\right) + \frac{1}{3}\right] \approx 1.1032$

13b. $\int_1^3 \frac{1}{x} dx \approx \frac{3-1}{3(8)}\left[1 + 4\left(\frac{4}{5}\right) + 2\left(\frac{2}{3}\right) + 4\left(\frac{4}{7}\right) + 2\left(\frac{1}{2}\right) + 4\left(\frac{4}{9}\right) + 2\left(\frac{2}{5}\right) + 4\left(\frac{4}{11}\right) + \frac{1}{3}\right] \approx 1.0987$

14a. $\Delta x = \frac{1-0}{4} = \frac{1}{4}$

$a = x_0 = 0$ $\qquad\quad f(x_0) = f(0) = 0$

$\qquad x_1 = 0.25$ $\qquad f(x_1) = f(0.25) = \frac{1}{256}$

$\qquad x_2 = 0.5$ $\qquad f(x_2) = f(0.5) = \frac{1}{16}$

$\qquad x_3 = 0.75$ $\qquad f(x_3) = f(0.75) = \frac{81}{256}$

$b = x_4 = 1$ $\qquad\quad f(x_4) = f(1) = 1$

$\int_0^1 x^4 dx \approx \frac{1-0}{2(4)}\left[0 + 2\left(\frac{1}{256}\right) + 2\left(\frac{1}{16}\right) + 2\left(\frac{81}{256}\right) + 1\right] \approx 0.2207$

14b. $\int_0^1 x^4 dx \approx \frac{1-0}{3(4)}\left[0 + 4\left(\frac{1}{256}\right) + 2\left(\frac{1}{16}\right) + 4\left(\frac{81}{256}\right) + 1\right] \approx 0.2005$

15a. $\Delta x = \frac{3-0}{6} = \frac{1}{2}$

$a = x_0 = 0$ $\qquad\quad f(x_0) = f(0) = 0$

$\qquad x_1 = 0.5$ $\qquad f(x_1) = f(0.5) = \sqrt[3]{0.5}$

$\qquad x_2 = 1$ $\qquad\quad f(x_2) = f(1) = 1$

$\qquad x_3 = 1.5$ $\qquad f(x_3) = f(1.5) = \sqrt[3]{1.5}$

$\qquad x_4 = 2$ $\qquad\quad f(x_4) = f(2) = \sqrt[3]{2}$

$\qquad x_5 = 2.5$ $\qquad f(x_5) = f(2.5) = \sqrt[3]{2.5}$

$b = x_6 = 3$ $\qquad\quad f(x_6) = f(3) = \sqrt[3]{3}$

$$\int_0^3 \sqrt[3]{x}\,dx \approx \frac{3-0}{2(6)}\left[0 + 2\sqrt[3]{0.5} + 2(1) + 2\sqrt[3]{1.5} + 2\sqrt[3]{2} + 2\sqrt[3]{2.5} + \sqrt[3]{3}\right] \approx 3.1383$$

15b. $$\int_0^3 \sqrt[3]{x}\,dx \approx \frac{3-0}{3(6)}\left[0 + 4\sqrt[3]{0.5} + 2(1) + 4\sqrt[3]{1.5} + 2\sqrt[3]{2} + 4\sqrt[3]{2.5} + \sqrt[3]{3}\right] \approx 3.1908$$

16a. $\Delta x = \dfrac{1-0}{4} = \dfrac{1}{4}$

$a = x_0 = 0 \qquad f(x_0) = f(0) = 0.5$

$x_1 = 0.25 \qquad f(x_1) = f(0.25) = \dfrac{1}{\sqrt{3.984375}}$

$x_2 = 0.5 \qquad f(x_2) = f(0.5) = \dfrac{1}{\sqrt{3.875}}$

$x_3 = 0.75 \qquad f(x_3) = f(0.75) = \dfrac{1}{\sqrt{3.578125}}$

$b = x_4 = 1 \qquad f(x_4) = f(1) = \dfrac{1}{\sqrt{3}}$

$$\int_0^1 \frac{1}{\sqrt{4-x^3}}\,dx \approx \frac{1-0}{2(4)}\left[0.5 + (2)\frac{1}{\sqrt{3.984375}} + (2)\frac{1}{\sqrt{3.875}} + (2)\frac{1}{\sqrt{3.578125}} + \frac{1}{\sqrt{3}}\right]$$
$$\approx 0.5191$$

16b. $$\int_0^1 \frac{1}{\sqrt{4-x^3}}\,dx \approx \frac{1-0}{3(4)}\left[0.5 + (4)\frac{1}{\sqrt{3.984375}} + (2)\frac{1}{\sqrt{3.875}} + (4)\frac{1}{\sqrt{3.578125}} + \frac{1}{\sqrt{3}}\right]$$
$$\approx 0.5177$$

17. $$\int_1^\infty 3x^{-2}\,dx = \lim_{b\to\infty}\left[\int_1^b 3x^{-2}\,dx\right]$$
$$= \lim_{b\to\infty}\left[3\frac{x^{-1}}{-1}\right]_1^b$$
$$= \lim_{b\to\infty}\left[\frac{-3}{b} - \frac{-3}{1}\right]$$
$$= 0 + 3 = 3$$

18. $$\int_1^\infty x^{-2/3}\,dx = \lim_{b\to\infty}\left[\int_1^b x^{-2/3}\,dx\right]$$
$$= \lim_{b\to\infty}\left[\frac{x^{1/3}}{\frac{1}{3}}\right]_1^b$$
$$= \lim_{b\to\infty}\left[3\sqrt[3]{b} - 3\right]$$

diverges

19.

$$\int_{-\infty}^{0} e^{4x}dx = \lim_{a \to -\infty} \left[\int_{a}^{0} e^{4x}dx \right]$$

$$= \lim_{a \to -\infty} \left[\frac{e^{4x}}{4} \right]_{a}^{0}$$

$$= \lim_{a \to -\infty} \left[\frac{1}{4} - \frac{e^{4a}}{4} \right]$$

$$= \frac{1}{4} - 0 = \frac{1}{4}$$

20.

$$\int_{1}^{\infty} x^{-3}dx = \lim_{b \to \infty} \left[\int_{1}^{b} x^{-3}dx \right]$$

$$= \lim_{b \to \infty} \left[\frac{x^{-2}}{-2} \right]_{1}^{b}$$

$$= \lim_{b \to \infty} \left[\frac{-1}{2b^2} - \frac{-1}{2} \right]$$

$$= 0 + \frac{1}{2} = \frac{1}{2}$$

21.

$$\int_{0}^{2} \frac{1}{(x-2)^2}dx = \lim_{b \to 2^-} \left[\int_{0}^{b} (x-2)^{-2}dx \right]$$

$$= \lim_{b \to 2^-} \left[\frac{(x-2)^{-1}}{-1} \right]_{0}^{b}$$

$$= \lim_{b \to 2^-} \left[\frac{-1}{b-2} - \frac{-1}{-2} \right]$$

diverges

22.

$$\int_{-\infty}^{0} xe^x dx = \lim_{a \to -\infty} \left[\int_{a}^{0} xe^x dx \right]$$

$$= \lim_{a \to -\infty} \left[xe^x - e^x \right]_{a}^{0} \quad \text{(Use formula 26.)}$$

$$= \lim_{a \to -\infty} \left[(0 - e^0) - (ae^a - e^a) \right]$$

$$= (-1) - (0) = -1 \quad \text{(See problem 32, section 7.4)}$$

23.

$$\int_{0}^{3} \sqrt{x^2 + 16}\,dx \qquad \text{Use formula 16 with } a = 4$$

$$= \frac{1}{2} \left[x\sqrt{x^2+16} + 16 + 16 \ln \left| x + \sqrt{x^2+16} \right| \right]_{0}^{3}$$

$$= \frac{1}{2} \left[3\sqrt{9+16} + 16 \ln \left| 3 + \sqrt{9+16} \right| \right] - \frac{1}{2} \left[0 + 16 \ln \left| 0 + \sqrt{16} \right| \right]$$

$$= \frac{1}{2}(3)(5) + 8 \ln 8 - 8 \ln 4 \approx 13.045$$

24.

$$\int -200xe^{-0.2x}dx$$

$u = -200x$	$dv = e^{-0.2x}$
$du = -200dx$	$v = \dfrac{e^{-0.2x}}{-0.2} = -5e^{-0.2x}$

$$= -200x(-5e^{-0.2x}) - \int (-5e^{-0.2x})(-200dx)$$

$$= 1000xe^{-0.2x} - 1000\int e^{-0.2x}dx$$

$$= 1000xe^{-0.2x} - 1000(-5e^{-0.2x}) + C$$

$$= 1000xe^{-0.2x} + 5000e^{-0.2x} + C$$

$$R(x) = \int (-200xe^{-0.2x} + 5e^{-0.2x})dx$$

$$= \left[1000xe^{-0.2x} + 5000e^{-0.2x}\right] + 5(-5e^{-0.2x}) + C$$

$$= 1000xe^{-0.2x} + 4975e^{-0.2x} + C$$

$$R(0) = 1000(0)e^0 + 4975e^0 + C$$

$$0 = 4975 + C$$

$$-4975 = C$$

$$R(x) = 1000xe^{-0.2x} + 4975e^{-0.2x} - 4975$$

Chapter 7 Test   (pages 466-467)

1.

$$\int x \ln 3x\, dx$$

$u = \ln 3x$	$dv = x dx$
$du = \dfrac{1}{3x}(3)dx$	$v = \dfrac{1}{2}x^2$
$= \dfrac{1}{x}dx$	

$$= \ln 3x\left(\tfrac{1}{2}x^2\right) - \int \tfrac{1}{2}x^2\left(\tfrac{1}{x}dx\right)$$

$$= \frac{x^2 \ln 3x}{2} - \frac{1}{2}\int x\, dx$$

$$= \frac{x^2 \ln 3x}{2} - \frac{1}{2}\frac{x^2}{2} + C$$

$$= \frac{x^2 \ln 3x}{2} - \frac{1}{4}x^2 + C \qquad \text{(b)}$$

2.

$$\Delta x = \frac{2-1}{4} = \frac{1}{4}$$

$$a = x_0 = 1$$

$$x_1 = \frac{5}{4}$$

$$x_2 = \frac{3}{2} \qquad \text{(a)}$$

3.

$u = x$	$dv = e^{-x}dx$
$du = dx$	$v = -e^{-x}$

$$\int xe^{-x}dx$$

$$= x(-e^{-x}) - \int (-e^{-x})dx$$

$$= -xe^{-x} - e^{-x} + C \qquad \text{(d)}$$

4. $\int (\ln x)^2 dx$ Use formula 23 with $n = 2$

$= x(\ln x)^2 - 2\int \ln x \, dx$ Use formula 22

$= x(\ln x)^2 - 2[x \ln x - x] + C$

$= x(\ln x)^2 - 2x \ln x + 2x + C$ (b)

5. $\int_1^\infty \frac{1}{x^2} dx = \lim_{a \to \infty} \left[\int_1^a x^{-2} dx \right]$

$= \lim_{a \to \infty} \left[-x^{-1} \right]_1^a$

$= \lim_{a \to \infty} \left[\frac{-1}{a} - \frac{-1}{1} \right]$

$= 0 + 1 = 1$ (a)

6. $\int_1^2 \ln x \, dx$ Use formula 22

$= x \ln x - x]_1^2$

$= [2 \ln 2 - 2] - [\ln 1 - 1]$

$= 2 \ln 2 - 2 + 1$

$= 2 \ln 2 - 1$ (c)

7. $\int_3^\infty \frac{1}{x^3} dx = \lim_{b \to \infty} \left[\int_3^b x^{-3} dx \right]$

$= \lim_{b \to \infty} \left[\frac{x^{-2}}{-2} \right]_3^b$

$= \lim_{b \to \infty} \left[\frac{-1}{2b^2} - \frac{-1}{2(9)} \right]$

$= 0 + \frac{1}{18} = \frac{1}{18}$ (a)

8. $\int x^2 \ln 3x \, dx$

$= \ln 3x \left(\frac{1}{3} x^3 \right) - \int \frac{1}{3} x^3 \left(\frac{1}{x} dx \right)$

$= \frac{x^3 \ln 3x}{3} - \frac{1}{3} \int x^2 \, dx$

$= \frac{x^3 \ln 3x}{3} - \frac{1}{3} \cdot \frac{x^3}{3} + C$

$= \frac{x^3 \ln 3x}{3} - \frac{x^3}{9} + C$

$u = \ln 3x$	$dv = x^2 dx$
$du = \frac{1}{3x}(3)dx$	$v = \frac{1}{3} x^3$
$= \frac{1}{x} dx$	

9. $\Delta x = \frac{2 - 0}{4} = \frac{1}{2}$

$a = x_0 = 0$ $f(x_0) = f(0) = 1$

$x_1 = 0.5$ $f(x_1) = f(0.5) = 1.125$

$x_2 = 1$ $f(x_2) = f(1) = 2$

$x_3 = 1.5$ $f(x_3) = f(1.5) = 4.375$

$b = x_4 = 2$ $f(x_4) = f(2) = 9$

$\int_0^2 (x^3 + 1)dx \approx \frac{2-0}{2(4)}[1 + 2(1.125) + 2(2) + 2(4.375) + 9 = 6.25$

10. $$\int_1^\infty \frac{4}{x^2}dx = \lim_{b\to\infty}\left[\int_1^b 4x^{-2}dx\right]$$

$$= \lim_{b\to\infty}\left[4\cdot\frac{x^{-1}}{-1}\right]_1^b$$

$$= \lim_{b\to\infty}\left[\frac{-4}{b} - \frac{-4}{1}\right]$$

$$= 0 + 4 = 4$$

11. $$\int xe^{-x}dx$$

$u = x$	$dv = e^{-x}dx$
$du = dx$	$v = -e^{-x}$

$$= x(-e^{-x}) - \int(-e^{-x})dx$$

$$= -xe^{-x} + \int e^{-x}dx$$

$$= -xe^{-x} - e^{-x} + C$$

$$C(x) = \int(5e^x - xe^{-x})dx$$

$$= 5e^x - [-xe^{-x} - e^{-x}] + C$$

$$= 5e^x + xe^{-x} + e^{-x} + C$$

$$C(0) = 5e^0 + 0 + e^0 + C$$

$$800 = 5 + 1 + C$$

$$794 = C$$

$$C(x) = 5e^x + xe^{-x} + e^{-x} + 794$$

CHAPTER 8: DIFFERENTIAL EQUATIONS

Section 8.1 (pages 474-476)

1. $\frac{dy}{dx} = 5$

$5 = 5$

3. $\frac{dy}{dx} = 5x^4$

$5x^4 = 5x^4$

5. $\frac{dy}{dx} = 2x$

$3x(2x) - 6y = 0$

$6x^2 - 6(x^2) = 0$

$0 = 0$

7. $\frac{dy}{dx} = \frac{1}{x}$

$x\left(\frac{1}{x}\right) - 1 = 0$

$1 - 1 = 0$

$0 = 0$

9. $\dfrac{dy}{dx} = 3e^{3x}$

$3e^{3x} - 3y + 15 = 0$

$3e^{3x} - 3(5 + e^{3x}) + 15 = 0$

$3e^{3x} - 15 - 3e^{3x} + 15 = 0$

$$0 = 0$$

11. $\dfrac{dy}{dx} = xe^x + e^x$

$x(xe^x + e^x) - xy - y = 0$

$x(xe^x + e^x) - x(xe^x) - xe^x = 0$

$x^2e^x + xe^x - x^2e^x - xe^x = 0$

$$0 = 0$$

13. $dy = 3dx$

$\displaystyle\int dy = \int 3dx$

$y = 3x + C$

15. $dy = (2x + 1)dx$

$\displaystyle\int dy = \int (2x + 1)dx$

$y = x^2 + x + C$

17. $dy = (6x^3 + x - 2)dx$

$\displaystyle\int dy = \int (6x^3 + x - 2)dx$

$y = \frac{3}{2}x^4 + \frac{1}{2}x^2 - 2x + C$

19. $dy = (x^{1/2} + 2)dx$

$\displaystyle\int dy = \int (x^{1/2} + 2)dx$

$y = \frac{2}{3}x^{3/2} + 2x + C$

$y = \frac{2}{3}\sqrt{x^3} + 2x + C$

21. $\dfrac{dy}{dx} = 4x$

$dy = 4xdx$

$\displaystyle\int dy = \int 4xdx$

$y = 2x^2 + C$

23. $dy = \frac{3}{2}x^2dx$

$\displaystyle\int dy = \int \frac{3}{2}x^2dx$

$y = \frac{1}{2}x^3 + C$

25. $x^2 \dfrac{dy}{dx} = -4$

$\dfrac{dy}{dx} = \dfrac{-4}{x^2}$

$dy = -4x^{-2}dx$

$\displaystyle\int dy = \int -4x^{-2}dx$

$y = 4x^{-1} + C$

$y = \frac{4}{x} + C$

27. $d^2y = 6xdx^2$

$\displaystyle\int d(dy) = \left[\int \int (6xdx)\right]dx$

$dy = (3x^2 + C_1)dx$

$\displaystyle\int dy = \int (3x^2 + C_1)dx$

$y = x^3 + C_1x + C_2$

29.
$$d^2y = 8e^{2x}dx^2$$
$$\int d(dy) = \left[\int \left(8e^{2x}dx\right)\right]dx$$
$$dy = \left(4e^{2x} + C_1\right)dx$$
$$\int dy = \int \left(4e^{2x} + C_1\right)dx$$
$$y = 2e^{2x} + C_1 x + C_2$$

31.
$$dy = 12xdx$$
$$\int dy = \int 12xdx$$
$$y = 6x^2 + C$$
$$8 = 6(1)^2 + C$$
$$8 = 6 + C$$
$$2 = C$$
$$y = 6x^2 + 2$$

33.
$$dy = (7 - 4x)dx$$
$$\int dy = \int (7 - 4x)dx$$
$$y = 7x - 2x^2 + C$$
$$3 = 0 - 0 + C$$
$$3 = C$$
$$y = 7x - 2x^2 + 3$$

35.
$$\frac{dy}{dx} = x - x^2$$
$$dy = (x - x^2)dx$$
$$\int dy = \int (x - x^2)dx$$
$$y = \tfrac{1}{2}x^2 - \tfrac{1}{3}x^3 + C$$
$$1 = \tfrac{1}{2} - \tfrac{1}{3} + C$$
$$\tfrac{5}{6} = C$$
$$y = \tfrac{1}{2}x^2 - \tfrac{1}{3}x^3 + \tfrac{5}{6}$$

37.
$$3\frac{dy}{dx} = 4x$$
$$dy = \tfrac{4}{3}xdx$$
$$\int dy = \int \tfrac{4}{3}xdx$$
$$y = \tfrac{2}{3}x^2 + C$$
$$-2 = 0 + C$$
$$-2 = C$$
$$y = \tfrac{2}{3}x^2 - 2$$

39.
$$x^2\frac{dy}{dx} = x^2 - 1$$
$$dy = \frac{x^2 - 1}{x^2}dx$$
$$\int dy = \int (1 - x^{-2})dx$$
$$y = x + x^{-1} + C$$
$$1 = 1 + 1 + C$$
$$-1 = C$$
$$y = x + \tfrac{1}{x} - 1$$

41.
$$x\frac{dy}{dx} = -3$$
$$dy = \frac{-3}{x}dx$$
$$\int dy = \int \frac{-3}{x}dx$$
$$y = -3 \ln |x| + C$$
$$4 = -3 \ln 1 + C$$
$$4 = C$$
$$y = -3 \ln |x| + 4$$

43.
$$dy = x^{1/3}dx$$
$$\int dy = \int x^{1/3}dx$$
$$y = \tfrac{3}{4}x^{4/3} + C$$
$$12 = \tfrac{3}{4}\left(\sqrt[3]{8}\right)^4 + C$$
$$12 = 12 + C$$
$$0 = C$$
$$y = \tfrac{3}{4}\sqrt[3]{x^4}$$

45.
$$d^2y = 2dx^2$$
$$\int d(dy) = \left[\int 2dx\right]dx$$
$$dy = (2x + C_1)dx$$
$$\frac{dy}{dx} = 2x + C_1$$
$$3 = 2(0) + C_1$$
$$3 = C_1$$
$$dy = (2x + 3)dx$$
$$\int dy = \int (2x + 3)dx$$
$$y = x^2 + 3x + C_2$$
$$5 = 0 + 0 + C_2$$
$$5 = C_2$$
$$y = x^2 + 3x + 5$$

47.
$$dC = (6 - 0.04x)dx$$
$$\int dC = \int (6 - 0.04x)dx$$
$$C = 6x - 0.02x^2 + k$$
$$400 = 6(100) - 0.02(100)^2 + k$$
$$400 = 600 - 200 + k$$
$$0 = k$$
$$C = 6x - 0.02x^2$$

49a.
$$dy = 600e^{3t}dt$$
$$\int dy = \int 600e^{3t}dt$$
$$y = 200e^{3t} + C$$

49b.
$$200 = 200e^0 + C$$
$$0 = C$$
$$y = 200e^{3t}$$

49c.
$$y = 200e^{15} \approx 6.538 \times 10^8$$

Section 8.2   (page 479-480)

1.
$$y\frac{dy}{dx} = 2x$$
$$ydy = 2xdx$$
$$\int ydy = \int 2xdx$$
$$\tfrac{1}{2}y^2 = x^2 + C_1$$
$$y^2 = 2x^2 + C$$

3.
$$\frac{dy}{y^2} = 2xdx$$
$$\int y^{-2}dy = \int 2xdx$$
$$-y^{-1} = x^2 + C_1$$
$$\tfrac{1}{y} = -x^2 + C$$

5.
$$x^2\frac{dy}{dx} = y^2$$
$$\frac{dy}{y^2} = \frac{dx}{x^2}$$
$$\int y^{-2}dy = \int x^{-2}dx$$
$$-y^{-1} = -x^{-1} + C$$
$$\tfrac{1}{y} = \tfrac{1}{x} + C$$

7.
$$\frac{dy}{dx} = 2y$$
$$\frac{dy}{y} = 2dx$$
$$\int \tfrac{1}{y}dy = \int 2dx$$
$$\ln |y| = 2x + C$$

9.

$$\frac{dy}{e^y} = x^2 dx$$

$$\int e^{-y} dy = \int x^2 dx$$

$$-e^{-y} = \frac{1}{3}x^3 + C_1$$

$$3e^{-y} = -x^3 + C$$

11.

$$\frac{dy}{y-1} = 2x dx$$

$$\int \frac{1}{y-1} dy = \int 2x dx$$

$$\ln |y-1| = x^2 + C$$

13.

$$\frac{dy}{dx} = (2x-1)y^2$$

$$\frac{dy}{y^2} = (2x-1)dx$$

$$\int y^{-2} dy = \int (2x-1)dx$$

$$-y^{-1} = x^2 - x + C_1$$

$$\frac{1}{y} = -x^2 + x + C$$

15.

$$y^2\frac{dy}{dx} = 2x$$

$$y^2 dy = 2x dx$$

$$\int y^2 dy = \int 2x dx$$

$$\frac{1}{3}y^3 = x^2 + C_1$$

$$y^3 = 3x^2 + C$$

$$(3)^3 = 0 + C$$

$$27 = C$$

$$y^3 = 3x^2 + 27$$

17.

$$x^2\frac{dy}{dx} = -y^2$$

$$\frac{dy}{-y^2} = \frac{dx}{x^2}$$

$$\int -y^{-2} dy = \int x^{-2} dx$$

$$y^{-1} = -x^{-1} + C$$

$$\frac{1}{y} = \frac{-1}{x} + C$$

$$1 = -1 + C$$

$$2 = C$$

$$\frac{1}{y} = \frac{-1}{x} + 2$$

19.

$$y dy = (x-4)dx$$

$$\int y dy = \int (x-4)dx$$

$$\frac{1}{2}y^2 = \frac{1}{2}x^2 - 4x + C_1$$

$$y^2 = x^2 - 8x + C$$

$$(-2)^2 = 0 - 0 + C$$

$$4 = C$$

$$y^2 = x^2 - 8x + 4$$

21.

$$2x\frac{dy}{dx} = x^2 y$$

$$\frac{dy}{dx} = \frac{x^2 y}{2x}$$

$$\frac{dy}{y} = \frac{1}{2}x dx$$

$$\int \frac{1}{y} dy = \int \frac{1}{2}x dx$$

$$\ln |y| = \frac{1}{4}x^2 + C$$

$$\ln |1| = \frac{1}{4}(4) + C$$

$$0 = 1 + C$$

$$-1 = C$$

$$\ln |y| = \frac{1}{4}x^2 - 1$$

$$4 \ln |y| = x^2 - 4$$

23. $y\dfrac{dy}{dx} = 4x$

$ydy = 4xdx$

$\displaystyle\int ydy = \int 4xdx$

$\frac{1}{2}y^2 = 2x^2 + C_1$

$y^2 = 4x^2 + C$

25. $y^2\dfrac{dy}{dx} = -e^{2x}$

$y^2dy = -e^{2x}dx$

$\displaystyle\int y^2dy = -\int e^{2x}dx$

$\frac{1}{3}y^3 = \frac{-1}{2}e^{2x} + C_1$

$2y^3 = -3e^{2x} + C$

27. $\dfrac{dy}{e^{2y}} = x^3dx$

$\displaystyle\int e^{-2y}dy = \int x^3dx$

$\frac{-1}{2}e^{-2y} = \frac{1}{4}x^4 + C_1$

$2e^{-2y} = -x^4 + C$

29. $\dfrac{dy}{y} = xe^{x^2}dx$

$\displaystyle\int \frac{1}{y}dy = \int xe^{x^2}dx$

$\displaystyle\int \frac{1}{y}dy = \frac{1}{2}\int e^{x^2}(2xdx)$

$\displaystyle\int \frac{1}{y}dy = \frac{1}{2}\int e^u du$

$\ln |y| = \frac{1}{2}e^u + C_1$

$\ln |y| = \frac{1}{2}e^{x^2} + C_1$

$2\ln |y| = e^{x^2} + C$

$\boxed{\begin{aligned} u &= x^2 \\ du &= 2xdx \end{aligned}}$

31. $ydy = (x^2 - 4)dx$

$\displaystyle\int ydy = \int (x^2 - 4)dx$

$\frac{1}{2}y^2 = \frac{1}{3}x^3 - 4x + C_1$

$3y^2 = 2x^3 - 24x + C$

$3(4)^2 = 0 - 0 + C$

$48 = C$

$3y^2 = 2x^3 - 24x + 48$

33. $\dfrac{dy}{y+3} = \dfrac{dx}{x^3}$

$\displaystyle\int \frac{1}{y+3}dy = \int x^{-3}dx$

$\ln |y+3| = \frac{x^{-2}}{-2} + C_1$

$2\ln |y+3| = \frac{-1}{x^2} + C$

$2\ln(1) = -1 + C$

$1 = C$

$2\ln |y+3| = \frac{-1}{x^2} + 1$

$\ln |y+3| = \frac{x^2 - 1}{2x^2}$

35. From problem 29: $\quad 2\ln |y| = e^{x^2} + C$

$2\ln e = e^0 + C$

$2 = 1 + C$

$1 = C$

$2\ln |y| = e^{x^2} + 1$

37.

$$P^{0.1}dP = \frac{-200}{e^{0.4t}}dt$$

$$\int P^{0.1}dP = \int -200e^{-0.4t}dt$$

$$\frac{P^{1.1}}{1.1} = -200\frac{e^{-0.4t}}{-0.4} + C_1$$

$$\frac{P^{1.1}}{1.1} = 500e^{-0.4t} + C_1$$

$$P^{1.1} = 550e^{-0.4t} + C$$

$$(38,000)^{1.1} = 550e^0 + C$$

$$C = (38,000)^{1.1} - 550$$

$$P^{1.1} = 550e^{-0.4t} + (38,000)^{1.1} - 550$$

$$P^{1.1} = 550e^{-0.4(10)} + (38,000)^{1.1} - 550$$

$$P^{1.1} \approx 108,544.3$$

$$P \approx 37,829$$

39.

$$\frac{dy}{(a-y)^2} = kdt$$

$$\int (a-y)^{-2}dy = \int kdt$$

$$(a-y)^{-1} = kt + C$$

$$\frac{1}{a-y} = kt + C$$

$$\frac{1}{kt+C} = a - y$$

$$\frac{1}{kt+C} - a = -y$$

$$y = a - \frac{1}{kt+C}$$

Section 8.3 (pages 486-488)

1.

$$\frac{dA}{dt} = kA$$

$$\frac{dA}{A} = kdt$$

$$\int \frac{1}{A}dA = \int kdt$$

$$\ln A = kt + C_1$$

$$A = e^{kt + C_1}$$

$$A = e^{C_1}e^{kt}$$

$$A = Ce^{kt}$$

3.

$$\frac{dA}{dt} = 0.06A$$

$$\frac{dA}{A} = 0.06dt$$

$$\int \frac{1}{A}dA = \int 0.06dt$$

$$\ln A = 0.06t + C_1$$

$$A = e^{0.06t + C_1}$$

$$A = e^{C_1}e^{0.06t}$$

$$A = Ce^{0.06t}$$

$$4000 = Ce^{0.06(0)}$$

$$4000 = C$$

$$A = 4000e^{0.06t}$$

$$A = 4000e^{0.06(10)}$$

$$A = \$7288.48$$

5. $\dfrac{dV}{dt} = kV$

$\dfrac{dV}{V} = kdt$

$\displaystyle\int \dfrac{1}{V}dV = \int kdt$

$\ln V = kt + C_1$

$V = e^{kt + C_1}$

$V = e^{C_1}e^{kt}$ so $V = Ce^{kt}$

When $t = 0$, $V = 3000$ When $t = 5$, $V = 10,000$ When $t = 10$

$3000 = Ce^{k(0)}$ $10,000 = 3000e^{k(5)}$ $V = 3000e^{2.4079456}$

$3000 = C$ $\dfrac{10}{3} = e^{5k}$ $= \$33{,}333$

$V = 3000e^{kt}$ $\ln\dfrac{10}{3} = 5k$

$\dfrac{\ln\dfrac{10}{3}}{5} = k$

$k \approx 0.24079456$

$V = 3000e^{0.24079456t}$

7. $\dfrac{dy}{dt} = ky(20,000 - y)$

$\dfrac{dy}{y(20,000 - y)} = kdt$

$\displaystyle\int \dfrac{1}{y(20,000 - y)}dy = \int kdt$ Use formula 4 with $a = 20,000$ and $b = -1$

$\dfrac{1}{20,000}\ln\left(\dfrac{y}{20,000 - y}\right) = k_1 t + C_1$

$\ln\left(\dfrac{y}{20,000 - y}\right) = 20,000(k_1 t + C_1)$

$\dfrac{y}{20,000 - y} = e^{20,000(k_1 t + C_1)}$

$\dfrac{y}{20,000 - y} = Ce^{kt}$

When $t = 0$, $y = 1000$ When $t = 2$, $y = 8000$

$\dfrac{1000}{20,000 - 1000} = Ce^{k(0)}$ $\dfrac{8000}{20,000 - 8000} = \dfrac{1}{19}e^{k(2)}$

$\dfrac{1}{19} = C$ $\dfrac{38}{3} = e^{2k}$

$\dfrac{y}{20,000 - y} = \dfrac{1}{19}e^{kt}$ $\ln\dfrac{38}{3} = 2k$

$\dfrac{\ln\dfrac{38}{3}}{2} = k$

$k \approx 1.27$

7. (continued)

$$\frac{y}{20,000 - y} = \frac{1}{19}e^{1.27t}$$

$$y = \frac{1}{19}e^{1.27t}(20,000 - y)$$

$$\frac{y}{\frac{1}{19}e^{1.27t}} = 20,000 - y$$

$$19e^{-1.27t}y = 20,000 - y$$

$$y + 19e^{-1.27t}y = 20,000$$

$$y(1 + 19e^{-1.27t}) = 20,000$$

$$y = \frac{20,000}{1 + 19e^{-1.27t}}$$

9. $\dfrac{dx}{dt} = k(4000 - x)$

11. $\dfrac{dy}{dt} = k(50,000 - y)$

$$\frac{dy}{50,000 - y} = kdt$$

$$\int \frac{1}{50,000 - y}dy = \int kdt$$

$$-\ln(50,000 - y) = kt + C_1$$

$$\ln(50,000 - y) = -kt - C_1$$

$$50,000 - y = e^{-kt - C_1}$$

$$50,000 - y = Ce^{-kt}$$

When $t = 0$, $y = 0$

$$50,000 = Ce^0$$

$$50,000 = C$$

$$50,000 - y = 50,000e^{-kt}$$

When $t = 10$, $y = 12,000$

$$50,000 - 12,000 = 50,000e^{-k(10)}$$

$$\frac{19}{25} = e^{-10k}$$

$$\frac{\ln\frac{19}{25}}{-10} = k$$

$$k \approx 0.027$$

$$50,000 - y = 50,000e^{-0.027t}$$

$$y = 50,000 - 50,000e^{-0.027t}$$

When $t = 30$

$$y = 50,000 - 50,000e^{-0.027(30)}$$

$$= 28,051$$

13. $\dfrac{dS}{dt} = kS^2$

15. $\dfrac{dQ}{dt} = kQ$

$\dfrac{dQ}{Q} = kdt$

$\displaystyle\int \dfrac{1}{Q}dQ = \int kdt$

$\ln Q = kt + C_1$

$Q = e^{kt + C_1}$

$Q = e^{C_1}e^{kt}$

$Q = Ce^{kt}$

When $t = 0$, $Q = 200$

$200 = Ce^0$

$200 = C$

$Q = 200e^{kt}$

When $t = 3$, $Q = 600$

$600 = 200e^{k(3)}$

$3 = e^{3k}$

$\dfrac{\ln 3}{3} = k$

$k \approx 0.37$

$Q = 200e^{0.37t}$

When $t = 5$

$Q = 200e^{0.37(5)}$

≈ 1272

17. $\dfrac{dR}{dt} = kR$

$\dfrac{dR}{R} = kdt$

$\displaystyle\int \dfrac{1}{R}dR = \int kdt$

$\ln R = kt + C_1$

$R = e^{kt + C_1}$

$R = e^{C_1}e^{kt}$

$R = Ce^{kt}$

19. $\dfrac{dT}{dt} = k(70 - T)$

$\dfrac{dT}{70 - T} = kdt$

$\displaystyle\int \dfrac{1}{70 - T}dT = \int kdt$

$-\ln |70 - T| = kt + C_1$

$\ln |70 - T| = -kt - C_1$

$|70 - T| = e^{-kt - C_1}$

$70 - T = Ce^{-kt}$

When $t = 0$, $T = 400$

$70 - 400 = Ce^0$

$-330 = C$

$70 - T = -330e^{-kt}$

When $t = 2$, $T = 300$

$70 - 300 = -330e^{-k(2)}$

$\dfrac{23}{33} = e^{-2k}$

$\dfrac{\ln \frac{23}{33}}{-2} = k$

$k \approx 0.18$

$70 - T = -330e^{-0.18t}$

$T = 70 + 330e^{-0.18t}$

21. $\dfrac{dT}{dt} = k(70 - T)$

From problem 19

$70 - T = C_1 e^{-kt}$

$-T = -70 + C_1 e^{-kt}$

$T = 70 + Ce^{-kt}$

Chapter 8 Review   (pages 489-490)

1. $dy = 4dx$

$$\int dy = \int 4dx$$

$$y = 4x + C$$

2. $dy = (3 - 4x)dx$

$$\int dy = \int (3 - 4x)dx$$

$$y = 3x - 2x^2 + C$$

3. $dy = (5x^2 + 3x - 2)dx$

$$\int dy = \int (5x^2 + 3x - 2)dx$$

$$y = \frac{5}{3}x^3 + \frac{3}{2}x^2 - 2x + C$$

4. $dy = (8x^4 - x^3 + 5)dx$

$$\int dy = \int (8x^4 - x^3 + 5)dx$$

$$y = \frac{8}{5}x^5 - \frac{1}{4}x^4 + 5x + C$$

5. $x^2 \frac{dy}{dx} = -6$

$$dy = \frac{-6}{x^2}dx$$

$$\int dy = \int (-6x^{-2})dx$$

$$y = 6x^{-1} + C$$

$$y = \frac{6}{x} + C$$

6. $dy = (4x + e^x)dx$

$$\int dy = \int (4x + e^x)dx$$

$$y = 2x^2 + e^x + C$$

7. $ydy = xdx$

$$\int ydy = \int xdx$$

$$\frac{1}{2}y^2 = \frac{1}{2}x^2 + C_1$$

$$y^2 = x^2 + C$$

8. $\frac{dy}{y^3} = xdx$

$$\int y^{-3}dy = \int xdx$$

$$\frac{y^{-2}}{-2} = \frac{x^2}{2} + C_1$$

$$\frac{1}{y^2} = -x^2 + C$$

9. $\frac{dy}{e^y} = xdx$

$$\int e^{-y}dy = \int xdx$$

$$-e^{-y} = \frac{1}{2}x^2 + C_1$$

$$2e^{-y} = -x^2 + C$$

10. $\frac{dy}{y^2} = e^x dx$

$$\int y^{-2}dy = \int e^x dx$$

$$-y^{-1} = e^x + C_1$$

$$\frac{1}{y} = -e^x + C$$

11.
$$dy = (5 - 3x)dx$$
$$\int dy = \int (5 - 3x)dx$$
$$y = 5x - \frac{3}{2}x^2 + C$$
$$0 = 5(2) - \frac{3}{2}(4) + C$$
$$0 = 10 - 6 + C$$
$$-4 = C$$
$$y = 5x - \frac{3}{2}x^2 - 4$$

12.
$$2\frac{dy}{dx} = 3x$$
$$dy = \frac{3}{2}xdx$$
$$\int dy = \int \frac{3}{2}xdx$$
$$y = \frac{3}{4}x^2 + C$$
$$7 = \frac{3}{4}(4) + C$$
$$4 = C$$
$$y = \frac{3}{4}x^2 + 4$$

13.
$$x^2\frac{dy}{dx} = x^2 - 2$$
$$\frac{dy}{dx} = \frac{x^2 - 2}{x^2}$$
$$dy = (1 - 2x^{-2})dx$$
$$\int dy = \int (1 - 2x^{-2})dx$$
$$y = x + 2x^{-1} + C$$
$$4 = 1 + 2 + C$$
$$1 = C$$
$$y = x + \frac{2}{x} + 1$$

14.
$$y^2dy = 2xdx$$
$$\int y^2dy = \int 2xdx$$
$$\frac{1}{3}y^3 = x^2 + C_1$$
$$y^3 = 3x^2 + C$$
$$(9)^3 = 3(0) + C$$
$$729 = C$$
$$y^3 = 3x^2 + 729$$

15.
$$\frac{dy}{y^2} = \frac{3}{x^2}dx$$
$$\int y^{-2}dy = \int 3x^{-2}dx$$
$$-y^{-1} = -3x^{-1} + C_1$$
$$\frac{1}{y} = \frac{3}{x} + C$$
$$\frac{1}{\left(\frac{1}{6}\right)} = \frac{3}{1} + C$$
$$6 = 3 + C$$
$$3 = C$$
$$\frac{1}{y} = \frac{3}{x} + 3$$

16.
$$3y^2\frac{dy}{dx} = 2x + 3$$
$$y^2dy = \left(\frac{2}{3}x + 1\right)dx$$
$$\int y^2dy = \int \left(\frac{2}{3}x + 1\right)dx$$
$$\frac{1}{3}y^3 = \frac{1}{3}x^2 + x + C_1$$
$$y^3 = x^2 + 3x + C$$
$$(8)^3 = 0 + 0 + C$$
$$512 = C$$
$$y^3 = x^2 + 3x + 512$$

17.

$$\frac{dy}{y+1} = \frac{dx}{x^2}$$

$$\int \frac{1}{y+1}dy = \int x^{-2}dx$$

$$\ln|y+1| = -x^{-1} + C_1$$

$$\ln|0+1| = \frac{-1}{\left(\frac{-1}{2}\right)} + C_1$$

$$0 = 2 + C_1$$

$$-2 = C_1$$

$$\ln|y+1| = \frac{-1}{x} - 2$$

18.

$$\frac{dy}{y} = xe^{x^2}dx$$

From problem 29, section 8.2

$$2\ln|y| = e^{x^2} + C$$

$$2\ln(1) = e^0 + C$$

$$0 = 1 + C$$

$$-1 = C$$

$$2\ln|y| = e^{x^2} - 1$$

19.

$$\frac{dA}{A} = 0.10dt$$

$$\int \frac{1}{A}dA = \int 0.10dt$$

$$\ln A = 0.10t + C_1$$

$$A = e^{0.10t + C_1}$$

$$A = e^{C_1} \cdot e^{0.10t}$$

$$A = Ce^{0.10}$$

When $t = 0$, $A = 2000$

$$2000 = Ce^0$$

$$2000 = C$$

$$A = 2000e^{0.10t}$$

20a.

$$\frac{dy}{dt} = k(50 - y)$$

20b.

$$\frac{dy}{50 - y} = kdt$$

$$\int \frac{1}{50 - y}dy = \int kdt$$

$$-\ln(50 - y) = kt + C_1$$

$$\ln(50 - y) = -kt - C_1$$

$$50 - y = e^{-kt - C_1}$$

$$50 - y = Ce^{-kt}$$

When $t = 0$, $y = 0$

$$50 - 0 = Ce^0$$

$$50 = C$$

$$50 - y = 50e^{-kt}$$

$$-y = -50 + 50e^{-kt}$$

$$y = 50 - 50e^{-kt}$$

20c. When $t = 10$, $y = 30$

$$30 = 50 - 50e^{-k(10)}$$

$$-20 = -50e^{-10k}$$

$$0.4 = e^{-10k}$$

$$\frac{\ln 0.4}{-10} = k$$

$$k \approx 0.092$$

$$y = 50 - 50e^{-0.092t}$$

When $t = 25$

$$y = 50 - 50e^{0.092(25)}$$

$$\approx 45$$

Chapter 8 Test (pages 490-491)

1.
$$dy = (6x^2 + 8x)dx$$
$$\int dy = \int (6x^2 + 8x)dx$$
$$y = 2x^3 + 4x^2 + C \quad \text{(b)}$$

2.
$$\frac{dy}{y} = (2x + 1)dx$$
$$\int \frac{1}{y}dy = \int (2x + 1)dx$$
$$\ln |y| = x^2 + x + C \quad \text{(d)}$$

3.
$$dy = (5 - 4x)dx$$
$$\int dy = \int (5 - 4x)dx$$
$$y = 5x - 2x^2 + C$$
$$10 = 5 - 2 + C$$
$$7 = C$$
$$y = 5x - 2x^2 + 7 \quad \text{(d)}$$

4.
$$x^3 y \frac{dy}{dx} = -1$$
$$ydy = \frac{-1}{x^3}dx$$
$$\int ydy = \int (-x^{-3})dx$$
$$\tfrac{1}{2}y^2 = \frac{x^{-2}}{2} + C_1$$
$$y^2 = \frac{1}{x^2} + C$$
$$(2)^2 = 1 + C$$
$$3 = C$$
$$y^2 = \frac{1}{x^2} + 3 \quad \text{(b)}$$

5. $\frac{dx}{dt} = k(8000 - x)$ \quad (b)

6.
$$dy = (2x - e^{-x})dx$$
$$\int dy = \int (2x - e^{-x})dx$$
$$y = x^2 + e^{-x} + C$$

7.
$$y^2 \frac{dy}{dx} = x + 2$$
$$y^2 dy = (x + 2)dx$$
$$\int y^2 dy = \int (x + 2)dx$$
$$\tfrac{1}{3}y^3 = \tfrac{1}{2}x^2 + 2x + C_1$$
$$2y^3 = 3x^2 + 12x + C$$

8.
$$dy = \tfrac{1}{2}(x + 1)^{-1/2}dx$$
$$\int dy = \int \tfrac{1}{2}(x + 1)^{-1/2}dx$$
$$y = (x + 1)^{1/2} + C$$
$$6 = \sqrt{0 + 1} + C$$
$$6 = 1 + C$$
$$5 = C$$
$$y = \sqrt{x + 1} + 5$$

9. $\dfrac{dA}{dt} = 0.09A$

$\dfrac{dA}{A} = 0.09dt$

$\displaystyle\int \dfrac{1}{A}dA = \int 0.09dt$

$\ln A = 0.09t + C_1$

$A = e^{0.09t + C_1}$

$A = Ce^{0.09t}$

When $t = 0$, $A = 5000$

$5000 = Ce^0$

$5000 = C$

$A = 5000e^{0.09t}$

$A = 5000e^{0.09(6)} \approx \$8,580.03$

10. $dQ = 800e^{0.2t}dt$

$\displaystyle\int dQ = \int 800e^{0.2t}dt$

$Q = 800\dfrac{e^{0.2t}}{0.2} + C$

$Q = 4000e^{0.2t} + C$

When $t = 0$, $Q = 4000$

$4000 = 4000e^{0.2(0)} + C$

$4000 = 4000 + C$

$0 = C$

$Q = 4000e^{0.2t}$

CHAPTER 9: MULTIVARIABLE CALCULUS

Section 9.1 (pages 495-497)

1. $f(0,1) = 2(0)^2 + 3(1)^2 - (0)(1) = 0 + 3 - 0 = 3$

3. $f(-1,2) = 2(-1)^2 + 3(2)^2 - (-1)(2) = 2 + 12 - (-2) = 16$

5. $f(0,0) = 0$

7. $f(-5,0) = 2(-5)^2 + 3(0) - (-5)(0) = 50 + 0 - 0 = 50$

9. $f(-5,-5) = 2(-5)^2 + 3(-5)^2 - (-5)(-5) = 50 + 75 - 25 = 100$

11. $g(3,4) = \sqrt{100 - (3)^2 - (4)^2} = \sqrt{100 - 9 - 16} = \sqrt{75} = \sqrt{25 \cdot 3} = 5\sqrt{3}$

13. $g(-6,8) = \sqrt{100 - (-6)^2 - (8)^2} = \sqrt{100 - 36 - 64} = \sqrt{0} = 0$

15. $g(0,2) = \sqrt{100 - 0 - (2)^2} = \sqrt{100 - 4} = \sqrt{96} = \sqrt{16 \cdot 6} = 4\sqrt{6}$

17. $f(3,4) = 2(3)^2 + 3(4)^2 - (3)(4) = 18 + 48 - 12 = 54$

19. $f(1,2) = 2(1)^2 + 3(2)^2 - (1)(2) = 2 + 12 - 2 = 12$

$g(6,8) = \sqrt{100 - (6)^2 - (8)^2} = \sqrt{100 - 36 - 64} = 0$

$f(1,2) - g(6,8) = 12 - 0 = 12$

21. $f(0,0,1) = -2(0) + 3(0) + (0)(0)(1) = 0$

23. $f(0,1,0) = -2(0) + 3(1)^2 + (0)(1)(0) = 0 + 3 + 0 = 3$

25. $f(-2,3,-1) = -2(-2)^2 + 3(3)^2 + (-2)(3)(-1) = -8 + 27 + 6 = 25$

27. $f(10,3,-1) = -2(10)^2 + 3(3)^2 + (10)(3)(-1) = -200 + 27 - 30 = -203$

29. $f(2,4,6) = -2(2)^2 + 3(4)^2 + (2)(4)(6) = -8 + 48 + 48 = 88$

31a. $R(50,60) = 300(50) + 420(60) = 15,000 + 25,200 = \$40,200$

31b. $R(45,65) = 300(45) + 420(65) = 13,500 + 27,300 = \$40,800$

33a. $D(6,30) = \dfrac{(6)(30)}{6+12} = \dfrac{180}{18} = 10$ 33b. $D(8,200) = \dfrac{(8)(200)}{8+12} = \dfrac{1600}{20} = 80$

35. $E(8,50,125) = 8(8)(50)(125) = 400,000$ foot-pounds

Section 9.2   (pages 503-504)

1,3,5.

7.

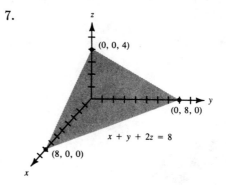

9.

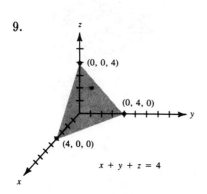

11.

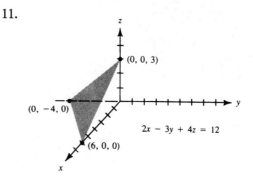

13. parallel to the y-axis

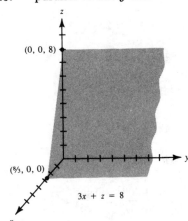

(0, 0, 8)

(8/3, 0, 0)

$3x + z = 8$

15. parallel to the xz plane

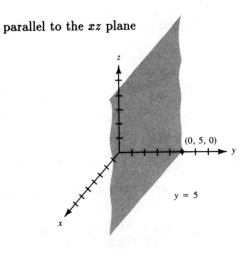

(0, 5, 0)

$y = 5$

17. $d = \sqrt{(0-3)^2 + (1-1)^2 + (4-2)^2} = \sqrt{9+0+4} = \sqrt{13}$

19. $d = \sqrt{(-1-2)^2 + (8-8)^2 + (9-5)^2} = \sqrt{9+0+16} = \sqrt{25} = 5$

21. center $(0,0,0)$
radius $\sqrt{25} = 5$

23. center $(0,1,-1)$
radius $\sqrt{16} = 4$

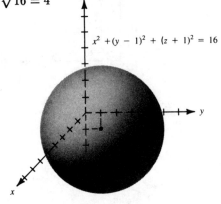

$x^2 + (y-1)^2 + (z+1)^2 = 16$

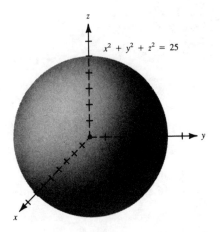

$x^2 + y^2 + z^2 = 25$

25. center $(0, 0, -2)$

radius $\sqrt{16} = 4$

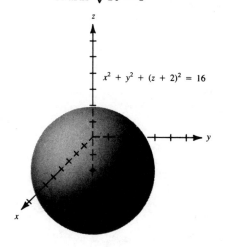

$x^2 + y^2 + (z + 2)^2 = 16$

27. $z = 0$

$(x - 2)^2 + y^2 + (0 - 3)^2 = 25$

$(x - 2)^2 + y^2 + 9 = 25$

$(x - 2)^2 + y^2 = 16$

circle with center $(2, 0, 0)$

and radius $\sqrt{16} = 4$

29. $z = 0$

$5x^2 - y^2 + 0^2 = 1$

$5x^2 - y^2 = 1$

hyperbola with center $(0, 0, 0)$

Section 9.3   (pages 509-513)

1a. $f_x = 2$

1b. $f_y = 3$

3a. $f_x = 1 + 12xy$

3b. $f_y = -1 + 6x^2$

5a. $f_x = 2e^{2x + 3y}$

5b. $f_y = 3e^{2x + 3y}$

7a. $f(x, y) = x^2 y^{1/2}$

$f_x = 2xy^{1/2} = 2x\sqrt{y}$

7b. $f_y = x^2\left(\frac{1}{2}y^{-1/2}\right) = \frac{x^2}{2y^{1/2}}$

9a. $f_x = \dfrac{4}{4x + 3y}$

9b. $f_y = \dfrac{3}{4x + 3y}$

11a. $f_x = \dfrac{2x}{x^2 + 2y^2}$

11b. $f_y = \dfrac{4y}{x^2 + 2y^2}$

13. $f_x(x, y) = 2xy$

$f_x(2, 4) = 2(2)(4)$

$= 16$

15. $f(x, y) = (x^2 + y^2)^{1/2}$

$\dfrac{\partial f}{\partial x} = \frac{1}{2}(x^2 + y^2)^{-1/2}(2x)$

$= \dfrac{x}{\sqrt{x^2 + y^2}}$

$\left.\dfrac{\partial f}{\partial x}\right|_{(3, 4)} = \dfrac{3}{\sqrt{(3)^2 + (4)^2}} = \dfrac{3}{\sqrt{25}} = \dfrac{3}{5}$

17. $\dfrac{\partial z}{\partial x} = \dfrac{y}{xy + 3y^2}$

$\left.\dfrac{\partial z}{\partial x}\right|_{(1,2)} = \dfrac{2}{(1)(2) + 3(2)^2}$

$= \dfrac{2}{2 + 12} = \dfrac{1}{7}$

19. $f_x(x,y) = 2e^{2x + 4y}$

$f_x(-2,1) = 2e^{2(-2) + 4(1)}$

$= 2e^0$

$= 2$

21. $f_x(x,y) = 3\sqrt{y}$

$f_x(3,\tfrac{1}{4}) = 3\sqrt{\tfrac{1}{4}}$

$= 3\left(\tfrac{1}{2}\right) = \tfrac{3}{2}$

23. $f_x = 10xy^3 - 12x$

$f_{xx} = 10y^3 - 12$

25. $f_y = 15x^2 y^2 + 14y$

$f_{yx} = 30xy^2$

27. $g_x = 6xe^{3x^2 + 2y}$

$g_{xx} = 6x\left[6xe^{3x^2 + 2y}\right] + 6e^{3x^2 + 2y}$

$= 6e^{3x^2 + 2y}[6x^2 + 1]$

29. $g_y = 2e^{3x^2 + 2y}$

$g_{yx} = 2 \cdot e^{3x^2 + 2y}(6x)$

$= 12xe^{3x^2 + 2y}$

31. $\dfrac{\partial z}{\partial x} = 4xy^3 - 5y^2$

$\dfrac{\partial^2 z}{\partial x^2} = 4y^3$

$\left.\dfrac{\partial^2 z}{\partial x^2}\right|_{(1,-2)} = 4(-2)^3 = -32$

33. $\dfrac{\partial z}{\partial y} = 6x^2 y^2 - 10xy$

$\dfrac{\partial^2 z}{\partial y^2} = 12x^2 y - 10x$

$\left.\dfrac{\partial^2 z}{\partial y^2}\right|_{(1,-2)} = 12(1)(-2) - 10(1)$

$= -24 - 10$

$= -34$

35. $f_x = 6e^{y^2}$

$f_{xx} = 0$

$f_{xy} = 6 \cdot 2ye^{y^2} = 12ye^{y^2}$

$f_y = 6x \cdot 2ye^{y^2} = 12xye^{y^2}$

$f_{yy} = 12xy\left[2ye^{y^2}\right] + 12xe^{y^2}$

$= 12xe^{y^2}[2y^2 + 1]$

$f_{yx} = 12ye^{y^2}$

37. $\dfrac{\partial z}{\partial x} = 10xy - ye^x$

$\dfrac{\partial^2 z}{\partial x^2} = 10y - ye^x$

$\dfrac{\partial^2 z}{\partial y \partial x} = 10x - e^x$

$\dfrac{\partial z}{\partial y} = 5x^2 - e^x$

$\dfrac{\partial^2 z}{\partial y^2} = 0$

$\dfrac{\partial^2 z}{\partial x \partial y} = 10x - e^x$

39a. $P(3,4) = 3(3)^2 + 2(4)^2$

$\qquad = 27 + 32$

$\qquad = 59 \qquad \$59,000$

39b. $\dfrac{\partial P}{\partial x} = 6x$

39c. $\left.\dfrac{\partial P}{\partial x}\right|_{(3,4)} = 6(3) = 18$

The profit increases by approximately $18,000 if the number of gas stoves is held constant at 400 and the number of electric stoves is increased by 100 $(x = 1)$.

41a. $\dfrac{\partial P}{\partial x} = 1000(0.5)x^{-0.5}y^{0.5}$

$\qquad = \dfrac{500y^{0.5}}{x^{0.5}}$

41b. $\dfrac{\partial P}{\partial y} = 1000x^{0.5}(0.5)y^{-0.5}$

$\qquad = \dfrac{500x^{0.5}}{y^{0.5}}$

43a. $\dfrac{\partial P}{\partial x} = 2000(0.25)x^{-0.75}y^{0.75}$

$\qquad = 500x^{-0.75}y^{0.75}$

$\dfrac{\partial^2 P}{\partial x^2} = 500(-0.75)x^{-1.75}y^{0.75}$

$\qquad = \dfrac{-375y^{0.75}}{x^{1.75}}$

43b. $\dfrac{\partial^2 P}{\partial y \partial x} = 500x^{-0.75}(0.75)y^{-0.25}$

$\qquad = \dfrac{375}{x^{0.75}y^{0.25}}$

45a. $\dfrac{\partial x}{\partial q} = 5$ The demand for the ranch model increases by approximately 5 units per dollar increase in selling price.

45b. $\dfrac{\partial y}{\partial p} = 3$ The demand for the colonial model increases by approximately 3 units per dollar increase in selling price.

47a. $\dfrac{\partial S}{\partial x} = 4y - 2x$

$\left.\dfrac{\partial S}{\partial x}\right|_{(2,5)} = 4(5) - 2(2)$

$\qquad = 20 - 4 = 16$

47b. $\left.\dfrac{\partial S}{\partial x}\right|_{(3,5)} = 4(5) - 2(3)$

$\qquad = 20 - 6 = 14$

This is less than the value found in part a.

49a. $P_d = -2 - x$

49b. $P_x = -3 - d$

51. $I_n = 0.03 + 0.5a$

Section 9.4 (pages 518-520)

1.
$$f_x = 6x - 2y + 4$$
$$f_y = 6y - 2x - 12$$
$$f_{xx} = 6$$
$$f_{yy} = 6$$
$$f_{xy} = -2$$

$$6x - 2y + 4 = 0$$
$$-2x + 6y - 12 = 0$$

$$18x - 6y + 12 = 0$$
$$-2x + 6y - 12 = 0$$
$$\overline{\hphantom{16x}}$$
$$16x \hspace{2em} = 0$$
$$x = 0$$
$$6(0) - 2y + 4 = 0$$
$$y = 2$$

$$K = (6)(6) - (-2)^2 = 36 - 4 = 32 \text{ and } f_{xx} = 6$$
$$f(0, 2) = -4 \text{ is a relative minimum.}$$

3.
$$f_x = 6x - 2y - 20$$
$$f_y = 6y - 2x - 4$$
$$f_{xx} = 6$$
$$f_{yy} = 6$$
$$f_{xy} = -2$$

$$6x - 2y - 20 = 0$$
$$-2x + 6y - 4 = 0$$

$$6x - 2y - 20 = 0$$
$$-6x + 18y - 12 = 0$$
$$\overline{\hphantom{16y}}$$
$$16y - 32 = 0$$
$$y = 2$$
$$6x - 2(2) - 20 = 0$$
$$x = 4$$

$$K = (6)(6) - (-2)^2 = 36 - 4 = 32 \text{ and } f_{xx} = 6$$
$$f(4, 2) = 4 \text{ is a relative minimum.}$$

5.
$$f_x = -6x - 2y + 4$$
$$f_y = -6y - 2x + 12$$
$$f_{xx} = -6$$
$$f_{yy} = -6$$
$$f_{xy} = -2$$

$$-6x - 2y + 4 = 0$$
$$-2x - 6y + 12 = 0$$

$$-6x - 2y + 4 = 0$$
$$6x + 18y - 36 = 0$$
$$\overline{\hphantom{16y}}$$
$$16y - 32 = 0$$
$$y = 2$$
$$-6x - 2(2) + 4 = 0$$
$$x = 0$$

$$K = (-6)(-6) - (-2)^2 = 36 - 4 = 32 \text{ and } f_{xx} = -6$$
$$f(0, 2) = -88 \text{ is a relative maximum.}$$

7.
$$f_x = -24x + 4y + 16$$
$$f_y = -6y + 4x - 8$$
$$f_{xx} = -24$$
$$f_{yy} = -6$$
$$f_{xy} = 4$$

$$-24x + 4y + 16 = 0$$
$$4x - 6y - 8 = 0$$

$$-24x + 4y + 16 = 0$$
$$24x - 36y - 48 = 0$$
$$\overline{\hphantom{-32y}}$$
$$-32y - 32 = 0$$
$$y = -1$$
$$4x - 6(-1) - 8 = 0$$
$$x = \tfrac{1}{2}$$

$$K = (-24)(-6) - (4)^2 = 144 - 16 = 128 \text{ and } f_{xx} = -24$$
$$f\left(\tfrac{1}{2}, -1\right) = 68 \text{ is a relative maximum.}$$

9.

$f_x = -2x + 10$ $-2x + 10 = 0$ $3y^2 - 6y - 9 = 0$

$f_y = 3y^2 - 6y - 9$ $x = 5$ $3(y^2 - 2y - 3) = 0$

$f_{xx} = -2$ $3(y + 1)(y - 3) = 0$

$f_{yy} = 6y - 6$ $y = -1$ $y = 3$

$f_{xy} = 0$

Test $(5, -1)$: $f_{yy}(5, -1) = 6(-1) - 6 = -12$

           $K = (-2)(-12) - (0)^2 = 24$ and $f_{xx} = -2$

           $f(5, -1) = -30$ is a relative maximum.

Test $(5, 3)$: $f_{yy}(5, 3) = 6(3) - 6 = 12$

           $K = (-2)(12) - 6 = -24 - 6 = -30$

           $(5, 3, -62)$ is a saddle point.

11.

$f_x = 12y - 3x^2$ $12x - 72y = 0$ $12y - 3x^2 = 0$

$f_y = 12x - 72y$ so $12x = 72y$ $12y - 3(6y)^2 = 0$

$f_{xx} = -6x$ $x = 6y$ $12y - 108y^2 = 0$

$f_{yy} = -72$ $12y(1 - 9y) = 0$

$f_{xy} = 12$ $y = 0$ $y = \dfrac{1}{9}$

           $x = 6(0)$ $x = 6\left(\dfrac{1}{9}\right)$

Test $(0, 0)$: $f_{xx}(0, 0) = -6(0) = 0$ $= 0$ $= \dfrac{2}{3}$

           $K = (0)(-72) - (12)^2 = -144$

           $(0, 0, 0)$ is a saddle point.

Test $\left(\dfrac{2}{3}, \dfrac{1}{9}\right)$: $f_{xx}\left(\dfrac{2}{3}, \dfrac{1}{9}\right) = -6\left(\dfrac{2}{3}\right) = -4$

           $K = (-4)(-72) - (12)^2 = 288 - 144 = 144$ and $f_{xx} = -4$

           $f\left(\dfrac{2}{3}, \dfrac{1}{9}\right) = \dfrac{4}{27}$ is a relative maximum.

13a.

$R_x = -10x + 39 - 2y$ $-10x - 2y + 39 = 0$ $-10x - 2y + 39 = 0$

$R_y = -16y + 39 - 2x$ $-2x - 16y + 39 = 0$ $\underline{10x + 80y - 195 = 0}$

$R_{xx} = -10$ $78y - 156 = 0$

$R_{yy} = -16$ $y = 2$

$R_{xy} = -2$ $-10x - 2(2) + 39 = 0$

           $x = \dfrac{7}{2}$

$K = (-10)(-16) - (-2)^2 = 160 - 4 = 156$ and $R_{xx} = -10$

R is maximized for $x = \dfrac{7}{2}$ and $y = 2$.

13b. $R(\frac{7}{2}, 2) = 127.25$ $127,250

15a. $C_x(0,0) = (0)^2 - 6(0) = 0$

$C_y(0,0) = -6(0) + 2(0) = 0$

Therefore $(0,0)$ is a critical value.

15b. $C_{xx}(0,0) = 2(0) = 0$

$K = (0)(2) - (-6)^2 = 0 - 36 = -36$

$(0,0,200)$ is a saddle point.

17.

$P_x = -4x + 6y$ $-4x + 6y = 0$ $-9y^2 + 6x = 0$

$P_y = -9y^2 + 6x$ so $-4x = -6y$ $-9y^2 + 6(\frac{3}{2}y) = 0$

$P_{xx} = -4$ $x = \frac{3}{2}y$ $-9y^2 + 9y = 0$

$P_{yy} = -18y$ $-9y(y-1) = 0$

$P_{xy} = 6$ $y = 0$ $y = 1$

$x = 0$ $x = \frac{3}{2}$

Test $(0,0)$: $P_{yy}(0,0) = -18(0) = 0$

$K = (-4)(0) - (6)^2 = -36$ (saddle point)

Test $\left(\frac{3}{2}, 1\right)$: $P_{yy}\left(\frac{3}{2}, 1\right) = -18\left(\frac{3}{2}\right) = -27$

$K = (-4)(-27) - (6)^2 = 108 - 36 = 72$ and $P_{xx} = -4$

Profit is a maximum for $x = \frac{3}{2}$ (150 mixers) and
$y = 1$ (100 blenders)

$P\left(\frac{3}{2}, 1\right) = 4.5$ ($4,500)

19. $lwh = 200$

$h = \frac{200}{lw}$

$C = 1.5(2l)\left(\frac{200}{lw}\right) + 1.5(2w)\left(\frac{200}{lw}\right) + 2(2lw)$

$= 600w^{-1} + 600l^{-1} + 4lw$

$C_l = -600l^{-2} + 4w$ $-600l^{-2} + 4w = 0$ $-600w^{-2} + 4l = 0$

$C_w = -600w^{-2} + 4l$ so $-150l^{-2} + w = 0$ $-600(150l^{-2})^{-2} + 4l = 0$

$C_{ll} = 1200l^{-3}$ $w = 150l^{-2}$ $\left(\frac{-600}{22500}\right)l^4 + 4l = 0$

$C_{ww} = 1200w^{-3}$ $\frac{-2}{75}l^4 + 4l = 0$

$C_{lw} = 4$ $l\left(\frac{-2}{75}l^3 + 4\right) = 0$

$l = 0$ $l^3 = 150$

$l = \sqrt[3]{150}$

Since l cannot be 0, use $l = \sqrt[3]{150}$:

$w = l = \sqrt[3]{150}$

$h = \dfrac{200}{\sqrt[3]{150}\,\sqrt[3]{150}} = \dfrac{200}{\sqrt[3]{150^2}}$

$C_{ll}(\sqrt[3]{150},\,\sqrt[3]{150}) = C_{ww}(\sqrt[3]{150},\,\sqrt[3]{150}) = \dfrac{1200}{150} = 8$

$K = (8)(8) - (4)^2 = 64 - 16 = 48$ and $C_{ll} = 8$

Cost is a minimum when $l = w = \sqrt[3]{150}$ and $h = \dfrac{200}{\sqrt[3]{150^2}}$.

Section 9.5 (pages 524-526)

1. $4x + y - 8 = 0$

$g(x, y) = 4x + y - 8$

$F(x, y, \lambda) = 2xy + \lambda(4x + y - 8)$

$\qquad\qquad = 2xy + 4x\lambda + y\lambda - 8\lambda$

$\begin{aligned} F_x &= 2y + 4\lambda = 0 & \lambda &= -2x \\ F_y &= 2x + \lambda = 0 & 2y + 4(-2x) &= 0 \\ F_\lambda &= 4x + y - 8 = 0 & 2y - 8x &= 0 \\ & & y &= 4x \\ & & 4x + (4x) - 8 &= 0 \\ & & x &= 1 \\ & & y &= 4(1) = 4 \\ & & \lambda &= -2 \end{aligned}$

$f(1, 4) = 8$ is a maximum.

3. $4x + 3y - 10 = 0$

$g(x, y) = 4x + 3y - 10$

$F(x, y, \lambda) = 4x^2 + 9y^2 + \lambda(4x + 3y - 10)$

$\qquad\qquad = 4x^2 + 9y^2 + 4x\lambda + 3y\lambda - 10\lambda$

$\begin{aligned} F_x &= 8x + 4\lambda = 0 & 4\lambda &= -8x \\ F_y &= 18y + 3\lambda = 0 & \lambda &= -2x \\ F_\lambda &= 4x + 3y - 10 = 0 & 18y + 3(-2x) &= 0 \\ & & 18y - 6x &= 0 \\ & & x &= 3y \\ & & 4(3y) + 3y - 10 &= 0 \\ & & 15y - 10 &= 0 \\ & & y &= \frac{10}{15} = \frac{2}{3} \\ & & x &= 3\left(\frac{2}{3}\right) = 2 \end{aligned}$

$f\left(2, \frac{2}{3}\right) = 20$ is a minimum. $\qquad\qquad \lambda = -2(2) = -4$

5. $4x + y - 26 = 0$

       $g(x,y) = 4x + y - 26$

       $F(x,y,\lambda) = 4x^2 + y^2 - 32y + 256 + \lambda(4x + y - 26)$

               $= 4x^2 + y^2 - 32y + 256 + 4x\lambda + y\lambda - 26\lambda$

$F_x = 8x + 4\lambda = 0$	$4\lambda = -8x$	$4x + y - 26 = 0$
$F_y = 2y - 32 + \lambda = 0$	$\lambda = -2x$	$-4x + 4y - 64 = 0$
$F_\lambda = 4x + y - 26 = 0$	$2y - 32 + (-2x) = 0$	$\overline{ 5y - 90 = 0}$
	$-2x + 2y - 32 = 0$	$y = 18$
		$4x + 18 - 26 = 0$
		$4x = 8$
		$x = 2$

       $f(2, 18) = 20$ is a minimum. $\lambda = -2(2) = -4$

7. $2x + 3y - 14 = 0$

       $g(x,y) = 2x + 3y - 14$

       $F(x,y,\lambda) = xy - 2x^2 + \lambda(2x + 3y - 14)$

               $= xy - 2x^2 + 2x\lambda + 3y\lambda - 14\lambda$

$F_x = y - 4x + 2\lambda = 0$	$x = -3\lambda$
$F_y = x + 3\lambda = 0$	$y - 4(-3\lambda) + 2\lambda = 0$
$F_\lambda = 2x + 3y - 14 = 0$	$y + 14\lambda = 0$
	$y = -14\lambda$
	$2(-3\lambda) + 3(-14\lambda) - 14 = 0$
	$-48\lambda = 14$
	$\lambda = \frac{-7}{24}$
	$x = -3\left(\frac{-7}{24}\right) = \frac{7}{8}$

       $f\left(\frac{7}{8}, \frac{49}{12}\right) = \frac{49}{24}$ is a maximum. $y = -14\left(\frac{-7}{24}\right) = \frac{49}{12}$

9. $x + 2y - 6 = 0$

$g(x,y) = x + 2y - 6$

$F(x,y,\lambda) = 4x^2 - 12y^2 + 4xy - 10x + 60y - 75 + \lambda(x + 2y - 6)$
$$= 4x^2 - 12y^2 + 4xy - 10x + 60y - 75 + x\lambda + 2y\lambda - 6\lambda$$

$F_x = 8x + 4y - 10 + \lambda = 0$ $\qquad\qquad\qquad -16x - 8y - 2\lambda = -20$

$F_y = -24y + 4x + 60 + 2\lambda = 0$ $\qquad\qquad\qquad \underline{4x - 24y + 2\lambda = -60}$

$F_\lambda = x + 2y - 6 = 0$ $\qquad\qquad\qquad\qquad -12x - 32y \qquad = -80$

$\qquad\qquad\qquad\qquad\qquad\qquad\qquad\qquad\qquad \underline{12x + 24y \qquad = 72}$

$\qquad\qquad\qquad\qquad\qquad\qquad\qquad\qquad\qquad\quad -8y \qquad = -8$

$\qquad\qquad\qquad\qquad\qquad\qquad\qquad\qquad\qquad\qquad\qquad y = 1$

$\qquad\qquad\qquad\qquad\qquad\qquad\qquad\qquad\qquad x + 2(1) - 6 = 0$

$\qquad\qquad\qquad\qquad\qquad\qquad\qquad\qquad\qquad\qquad\qquad x = 4$

$\qquad\qquad\qquad\qquad\qquad\qquad\qquad\qquad 8(4) + 4(1) - 10 + \lambda = 0$

$f(4,1) = 13$ is a maximum. $\qquad\qquad\qquad\qquad\qquad\qquad \lambda = -26$

11. $f(x,y) = xy$

$x + y = 124$

$g(x,y) = x + y - 124$

$F(x,y,\lambda) = xy + \lambda(x + y - 124)$
$$= xy + x\lambda + y\lambda - 124\lambda$$

$F_x = y + \lambda = 0$ $\qquad\qquad\qquad y = -\lambda$

$F_y = x + \lambda = 0$ $\qquad\qquad\qquad x = -\lambda$

$F_\lambda = x + y - 124 = 0$ $\qquad\qquad$ so $x = y$ $\qquad\qquad x + x - 124 = 0$

$\qquad\qquad\qquad\qquad\qquad\qquad\qquad\qquad\qquad\qquad\qquad 2x = 124$

$\qquad\qquad\qquad\qquad\qquad\qquad\qquad\qquad\qquad\qquad\qquad x = 62$

The product is $(62)(62) = 3844$. $\qquad\qquad\qquad\qquad y = 62$

13. $f(x,y) = xy$

$2x + 2y = 80$ $\qquad$ so $x + y = 40$

$g(x,y) = x + y - 40$

$F(x,y,\lambda) = xy + \lambda(x + y - 40)$
$$= xy + x\lambda + y\lambda - 40\lambda$$

$F_x = y + \lambda = 0$ $\qquad\qquad\qquad y = -\lambda$

$F_y = x + \lambda = 0$ $\qquad\qquad\qquad x = -\lambda$

$F_\lambda = x + y - 40 = 0$ $\qquad\qquad$ so $x = y$ $\qquad\qquad x + x - 40 = 0$

$\qquad\qquad\qquad\qquad\qquad\qquad\qquad\qquad\qquad\qquad\qquad 2x = 40$

$\qquad\qquad\qquad\qquad\qquad\qquad\qquad\qquad\qquad\qquad\qquad x = 20$

Maximum area is $(20)(20) = 400$ cm^2 . $\qquad\qquad\qquad y = 20$

15.   length $= y$ perimeter of a cross section $= 2(2x) + 2(x) = 6x$

$6x + y = 90$

$g(x, y) = 6x + y - 90$

Maximize the volume: $f(x, y) = y(2x)x = 2x^2 y$

$F(x, y, \lambda) = 2x^2 y + \lambda(6x + y - 90)$

$\qquad = 2x^2 y + 6x\lambda + y\lambda + y\lambda - 90\lambda$

$F_x = 4xy + 6\lambda = 0$ $\lambda = -2x^2$

$F_y = 2x^2 + \lambda = 0$ $4xy + 6(-2x^2) = 0$

$F_\lambda = 6x + y - 90 = 0$ $4xy - 12x^2 = 0$

$\qquad\qquad\qquad\qquad\qquad\qquad 4x(y - 3x) = 0$

$\qquad\qquad\qquad\qquad 4x = 0 \qquad y = 3x \qquad$ Since $x \neq 0$ use $y = 3x$.

$\qquad\qquad\qquad\qquad\qquad\qquad 6x + y - 90 = 0$

$\qquad\qquad\qquad\qquad\qquad\qquad 6x + 3x - 90 = 0$

$\qquad\qquad\qquad\qquad\qquad\qquad\qquad\qquad 9x = 90$

$\qquad\qquad\qquad\qquad\qquad\qquad\qquad\qquad\quad x = 10$

$\qquad\qquad\qquad\qquad\qquad\qquad\qquad\; y = 3(10) = 30$

Volume is a maximum for length 30 in., $2x = 2(10) = 20$
width 20 in. and height 10 in.

17.   length $=$ width $= x$ height $= y$

Minimize the area of the top and bottom, and the four sides:

$f(x, y) = 2x^2 + 4xy$

Volume $= 64$

$\qquad x^2 y = 64$

$g(x, y) = x^2 y - 64$

$F(x, y, \lambda) = 2x^2 + 4xy + \lambda(x^2 y - 64)$

$\qquad = 2x^2 + 4xy + x^2 y\lambda - 64\lambda$

$F_x = 4x + 4y + 2xy\lambda = 0$ $4x + x^2\lambda = 0$

$F_y = 4x + x^2\lambda = 0$ $x(4 + x\lambda) = 0$

$F_\lambda = x^2 y - 64 = 0$ $x = 0 \qquad x\lambda = -4$

$\qquad\qquad\qquad\qquad\qquad\qquad\qquad\qquad\qquad \lambda = \frac{-4}{x}$

Since $x \neq 0$ use $\lambda = \frac{-4}{x}$: $4x + 4y + 2xy\left(\frac{-4}{x}\right) = 0$, $4x + 4y - 8y = 0$, $4x = 4y$, $x = y$

$\qquad\qquad\qquad\qquad\qquad\qquad x^2(x) - 64 = 0$

$\qquad\qquad\qquad\qquad\qquad\qquad\qquad\; x^3 = 64$

$\qquad\qquad\qquad\qquad\qquad\qquad\qquad\; x = 4 \qquad y = 4$

The dimensions should be 4 in. $\times$ 4 in. $\times$ 4 in.

19.
$$f(x,y) = 2\pi x^2 + 2\pi xy$$
$$\pi x^2 y = 345$$
$$g(x,y) = \pi x^2 y - 345$$
$$F(x,y,\lambda) = 2\pi x^2 + 2\pi xy + \lambda(\pi x^2 y - 345)$$
$$= 2\pi x^2 + 2\pi xy + \pi x^2 y\lambda - 345\lambda$$

$F_x = 4\pi x + 2\pi y + 2\pi xy\lambda = 0$ $\qquad\qquad$ $2\pi x + \pi x^2\lambda = 0$

$F_y = 2\pi x + \pi x^2\lambda = 0$ $\qquad\qquad\qquad$ $\pi x(2 + x\lambda) = 0$

$F_\lambda = \pi x^2 y - 345 = 0$ $\qquad\qquad\quad$ $x = 0 \qquad x\lambda = -2$

$$\lambda = \tfrac{-2}{x}$$

Since $x \neq 0$ use $\lambda = \tfrac{-2}{x}$: $\qquad 4\pi x + 2\pi y + 2\pi xy\left(\tfrac{-2}{x}\right) = 0$

$$4\pi x + 2\pi y - 4\pi y = 0$$
$$4\pi x = 2\pi y$$
$$2x = y$$

$$\pi x^2(2x) - 345 = 0$$
$$2\pi x^3 = 345$$
$$x^3 = \frac{345}{2\pi}$$
$$x \approx 3.8$$
$$y = 2(3.8) = 7.6$$

The surface area is a minimum for a radius of 3.8 cm and a height of 7.6 cm.

21. $\quad x$ = length of short sides
y = length of long sides

Maximize the area. $f(x,y) = xy$

To fence 3 sides: $2x + y = 200$

$$g(x,y) = 2x + y - 200$$
$$F(x,y,\lambda) = xy + \lambda(2x + y - 200)$$
$$= xy + 2x\lambda + y\lambda - 200\lambda$$

$F_x = y + 2\lambda = 0$ $\qquad\qquad\qquad\qquad\qquad$ $\lambda = -x$

$F_y = x + \lambda = 0$ $\qquad\qquad\qquad\qquad\qquad$ $y + 2(-x) = 0$

$F_\lambda = 2x + y - 200 = 0$ $\qquad\qquad\qquad\qquad$ $y = 2x$

$$2x + (2x) - 200 = 0$$
$$4x = 200$$
$$x = 50 \qquad y = 2(50) = 100$$

The dimensions should be 50 ft. × 100 ft.

23. $x = \text{length} = \text{width}$
$y = \text{height}$

$\text{cost} = \text{cost of top} + \text{cost of bottom} + \text{cost of 4 sides}$

$f(x,y) = 0.10x^2 + 0.05x^2 + 0.05(4xy)$

$\qquad = 0.15x^2 + 0.20xy$

$\text{volume} = 96 \text{ in}^3$

$\qquad x^2y = 96$

$g(x,y) = x^2y - 96$

$F(x,y,\lambda) = 0.15x^2 + 0.20xy + \lambda(x^2y - 96)$

$\qquad = 0.15x^2 + 0.20xy + x^2y\lambda - 96\lambda$

$F_x = 0.30x + 0.20y + 2xy\lambda = 0$ $0.20x + x^2\lambda = 0$

$F_y = 0.20x + x^2\lambda = 0$ $x(0.20 + x\lambda) = 0$

$F_\lambda = x^2y - 96 = 0$ $x = 0$ $x\lambda = -0.20$

$\qquad\qquad\qquad\qquad\qquad\qquad\qquad\qquad\qquad \lambda = \frac{-0.20}{x}$

Since $x \neq 0$, use $\lambda = \frac{-0.20}{x}$:

$\qquad 0.30x + 0.20y + 2xy\left(\frac{-0.20}{x}\right) = 0$

$\qquad\qquad 0.30x + 0.20y - 0.40y = 0$

$\qquad\qquad\qquad\qquad 0.30x = 0.20y$

$\qquad\qquad\qquad\qquad \frac{3x}{2} = y$

$\qquad\qquad x^2\left(\frac{3x}{2}\right) - 96 = 0$

$\qquad\qquad\qquad\qquad \frac{3x^3}{2} = 96$

$\qquad\qquad\qquad\qquad x^3 = 64$

$\qquad\qquad\qquad x = 4$ $y = \frac{3(4)}{2} = 6$

The most economical box is 4 in. × 4 in. × 6 in.

Section 9.6   (pages 534-537)

1. $d_1 = 4 - (-2m + b) = 4 + 2m - b$

$d_2 = 10 - (0 + b) = 10 - b$

$d_3 = 12 - (2m + b) = 12 - 2m - b$

$f(m,b) = (4 + 2m - b)^2 + (10 - b)^2 + (12 - 2m - b)^2$

$f_m(m,b) = 2(4 + 2m - b)(2) + 0 + 2(12 - 2m - b)(-2)$

$\qquad = 16 + 8m - 4b - 48 + 8m + 4b$

$\qquad = -32 + 16m$

$$f_b(m,b) = 2(4+2m-b)(-1) + 2(10-b)(-1) + 2(12-2m-b)(-1)$$
$$= -8-4m+2b-20+2b-24+4m+2b$$
$$= -52+6b$$

Setting $f_m = 0$ and $f_b = 0$:

$$-32+16m = 0 \qquad\qquad -52+6b = 0$$
$$m = 2 \qquad\qquad b = \frac{52}{6} = \frac{26}{3}$$
$$y = 2x + \frac{26}{3}$$

3. $$d_1 = 10-(0+b) = 10-b$$
$$d_2 = 6-(2m+b) = 6-2m-b$$
$$d_3 = 5-(3m+b) = 5-3m-b$$
$$f(m,b) = (10-b)^2 + (6-2m-b)^2 + (5-3m-b)^2$$
$$f_m(m,b) = 0 + 2(6-2m-b)(-2) + 2(5-3m-b)(-3)$$
$$= -24+8m+4b-30+18m+6b$$
$$= -54+26m+10b$$
$$f_b(m,b) = 2(10-b)(-1) + 2(6-2m-b)(-1) + 2(5-3m-b)(-1)$$
$$= -20+2b-12+4m+2b-10+6m+2b$$
$$= -42+10m+6b$$

Setting $f_m = 0$ and $f_b = 0$:

$$26m+10b = 54 \qquad 13m+5b = 27 \qquad 39m+15b = 81$$
$$10m+6b = 42 \qquad 5m+3b = 21 \qquad \underline{-25m-15b = -105}$$
$$14m \qquad\quad = -24$$
$$m = \frac{-24}{14} = \frac{-12}{7}$$

To find b: $\quad 5m+3b = 21$
$$5\left(\frac{-12}{7}\right) + 3b = 21$$
$$3b = 21 + \frac{60}{7}$$
$$b = \frac{1}{3}\left(\frac{207}{7}\right) = \frac{69}{7}$$
$$y = \frac{-12}{7}x + \frac{69}{7}$$

5. $$d_1 = -11-(-4m+b) = -11+4m-b$$
$$d_2 = -3-(0+b) = -3-b$$
$$d_3 = 6-(4m+b) = 6-4m-b$$

$$f(m, b) = (-11 + 4m - b)^2 + (-3 - b)^2 + (6 - 4m - b)^2$$

$$f_m(m, b) = 2(-11 + 4m - b)(4) + 0 + 2(6 - 4m - b)(-4)$$

$$= -88 + 32m - 8b - 48 + 32m + 8b$$

$$= 136 + 64m$$

$$f_b(m, b) = 2(-11 + 4m - b)(-1) + 2(-3 - b)(-1) + 2(6 - 4m - b)(-1)$$

$$= 22 - 8m + 2b + 6 + 2b - 12 + 8m + 2b$$

$$= 16 + 6b$$

Setting $f_m = 0$ and $f_b = 0$:

$$136 + 64m = 0 \qquad\qquad 16 + 6b = 0$$

$$64m = 136 \qquad\qquad 6b = -16$$

$$m = \frac{136}{64} = \frac{17}{8} \qquad\qquad b = \frac{-16}{6} = \frac{-8}{3}$$

$$y = \frac{17}{8}x - \frac{8}{3}$$

7. $$d_1 = 5 - (0 + b) = 5 - b$$

$$d_2 = 6 - (m + b) = 6 - m - b$$

$$d_3 = 7 - (2m + b) = 7 - 2m - b$$

$$d_4 = 8 - (3m + b) = 8 - 3m - b$$

$$d_5 = 10 - (4m + b) = 10 - 4m - b$$

$$f(m, b) = (5 - b)^2 + (6 - m - b)^2 + (7 - 2m - b)^2 + (8 - 3m - b)^2 + (10 - 4m - b)^2$$

$$f_m(m, b) = 0 + 2(6 - m - b)(-1) + 2(7 - 2m - b)(-2) + 2(8 - 3m - b)(-3)$$

$$+ 2(10 - 4m - b)(-4)$$

$$= -12 + 2m + 2b - 28 + 8m + 4b - 48 + 18m + 6b - 80 + 32m + 8b$$

$$= -168 + 60m + 20b$$

$$f_b(m, b) = 2(5 - b)(-1) + 2(6 - m - b)(-1) + 2(7 - 2m - b)(-1)$$

$$+ 2(8 - 3m - b)(-1) + 2(10 - 4m - b)(-1)$$

$$= -10 + 2b - 12 + 2m + 2b - 14 + 4m + 2b - 16 + 6m + 2b - 20$$

$$+ 8m + 2b$$

$$= -72 + 20m + 10b$$

Setting $f_m = 0$ and $f_b = 0$:

$$60m + 20b = 168 \qquad\qquad 60m + 20b = 168$$

$$20m + 10b = 72 \qquad\qquad -60m - 30b = -216$$

$$-10b = -48 \qquad b = \frac{-48}{10} = \frac{24}{5}$$

To find m:

$$20m + 10\left(\frac{24}{5}\right) = 72$$

$$20m = 24$$

$$m = \frac{24}{20} = \frac{6}{5}$$

$$y = \frac{6}{5}x + \frac{24}{5}$$

9.
$$y = 2(3) + \frac{26}{3}$$

$$= 6 + \frac{26}{3} = \frac{44}{3}$$

11.
$$y = \frac{-12}{7}(4) + \frac{69}{7}$$

$$= \frac{21}{7} = 3$$

13.
$$y = \frac{17}{8}(6) - \frac{8}{3}$$

$$= \frac{51}{4} - \frac{8}{3}$$

$$= \frac{153}{12} - \frac{32}{12} = \frac{121}{12}$$

15.
$$y = \frac{6}{5}(10) + \frac{24}{5}$$

$$= 12 + \frac{24}{5}$$

$$= \frac{84}{5}$$

17.

x	y	x^2	xy
-2	7	4	-14
0	11	0	0
2	16	4	32
sum 0	34	8	18

$$m = \frac{3(18) - (0)(34)}{3(8) - (0)^2}$$

$$= \frac{54}{24} = \frac{9}{4}$$

$$b = \frac{(34)(8) - (0)(18)}{3(8) - (0)^2}$$

$$= \frac{34}{3}$$

$$y = \frac{9}{4}x + \frac{34}{3}$$

19.

x	y	x^2	xy
-2	14	4	-28
-1	11	1	-11
0	10	0	0
1	9	1	9
2	6	4	12
sum 0	50	10	-18

$$m = \frac{5(-18) - (0)(50)}{5(10) - (0)^2}$$

$$= \frac{-90}{50} = \frac{-9}{5}$$

$$b = \frac{(50)(10) - (0)(-18)}{5(10) - (0)^2}$$

$$= \frac{500}{50} = 10$$

$$y = \frac{-9}{5}x + 10$$

21.

x	y	x^2	xy
1	2	1	2
2	7	4	14
3	8	9	24
4	10	16	40
5	14	25	70
sum 15	41	55	150

$$m = \frac{5(150) - (15)(41)}{5(55) - (15)^2}$$

$$= \frac{135}{50} = 2.7$$

$$b = \frac{(41)(55) - (15)(150)}{5(55) - (15)^2}$$

$$= \frac{5}{50} = 0.1$$

$$y = 2.7x - 0.1$$

23a. Let $x = 0$ for 1975, $x = -1$ for 1970 and $x = 1$ for 1980.

x	y	x^2	xy
-1	10.6	1	-10.6
0	15.8	0	0
1	23.1	1	23.1
sum 0	49.5	2	12.5

$$m = \frac{3(12.5) - (0)(49.5)}{3(2) - (0)^2}$$

$$= \frac{12.5}{2} = 6.25$$

$$b = \frac{(49.5)(2) - (0)(12.5)}{3(2) - (0)^2}$$

$$= \frac{49.5}{3} = 16.5$$

$$y = 6.25x + 16.5$$

23b. Let $x = 4$ for 1995.

$$y = 6.25(4) + 16.5 = 41.5$$

25a. Let $x = 0$ for 1948.

x	y	x^2	xy
0	0.86	0	0
15	1.65	225	24.75
33	4.95	1089	163.35
34	4.95	1156	168.30
sum 82	12.41	2470	356.40

$$m = \frac{4(356.4) - (82)(12.41)}{4(2470) - (82)^2}$$

$$= \frac{407.98}{3156} \approx 0.12927$$

$$b = \frac{(12.41)(2470) - (82)(356.40)}{4(2470) - (82)^2}$$

$$= \frac{1427.9}{3156} \approx 0.45244$$

$$y = 0.12927x + 0.45244$$

25b. $y = 0.12927(45) + 0.45244 \approx \$6.27$

25c. No, the stamps are only worth \$6.27.

27.

x	y	x^2	xy
28	7.5	784	210
42	5.7	1764	239.4
51	4.9	2601	249.9
sum 121	18.1	5149	699.3

$$m = \frac{3(699.3) - (121)(18.1)}{3(5149) - (121)^2}$$

$$= \frac{-92.2}{806} \approx -0.11439$$

$$b = \frac{(18.1)(5149) - (121)(699.3)}{3(5149) - (121)^2}$$

$$= \frac{8581.6}{806} \approx 10.6471$$

$$y = -0.11439x + 10.6471$$

29. Multiply the first equation by 3 and the second equation by -13:

$$546m + 126b = 747.6$$
$$\underline{-546m - 156b = -743.6}$$
$$-30b = 4$$

$$b = \frac{-4}{30} = \frac{-2}{15}$$

$$42m + 12\left(\frac{-2}{15}\right) = 57.2$$

$$630m - 24 = 858$$

$$630m = 882$$

$$m = \frac{882}{630} = \frac{7}{5}$$

Chapter 9 Review (pages 537-538)

1. $f(2,-1) = 3(2)^2 + 2(2)^2(-1)^3 - (2)(-1)^2 = 12 - 8 - 2 = 2$

2.

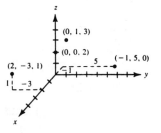

3.

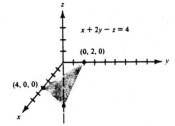

4. $d = \sqrt{(0-3)^2 + (1-1)^2 + (6-4)^2} = \sqrt{9 + 0 + 4} = \sqrt{13}$

5. center $(0,0,0)$
radius $\sqrt{64} = 8$

6. $f_x = ye^{xy} + 12x^2y^2$

7. $f_y = xe^{xy} + 8x^3y - 12y^3$

8. $f_{xx} = y \cdot ye^{xy} + 24xy^2$

$\quad = y^2 e^{xy} + 24xy^2$

9. $f_{xy} = [y(xe^{xy}) + e^{xy}(1)] + 24x^2 y$

$\quad = xye^{xy} + e^{xy} + 24x^2 y$

10. $f_{yx} = [x(ye^{xy}) + e^{xy}(1)] + 24x^2 y$

$\quad = xye^{xy} + e^{xy} + 24x^2 y$

11. $f_{yy} = x \cdot xe^{xy} + 8x^3 - 36y^2$

$\quad = x^2 e^{xy} + 8x^3 - 36y^2$

12. $f_x = 6x + 4y - 8$ $\qquad$ $6x + 4y - 8 = 0$ $\qquad$ $18x + 12y - 24 = 0$

 $f_y = 6y + 4x + 4$ $\qquad$ $4x + 6y + 4 = 0$ $\qquad$ $\underline{-8x - 12y \ - \ 8 = 0}$

 $f_{xx} = 6$ $\hspace{6.5cm}$ $10x \hspace{1.4cm} - 32 = 0$

 $f_{yy} = 6$ $\hspace{6.8cm}$ $x = \dfrac{32}{10} = \dfrac{16}{5}$

 $f_{xy} = 4$ $\hspace{5.8cm}$ $4\left(\dfrac{16}{5}\right) + 6y + 4 = 0$

$\hspace{8.5cm}$ $6y = -4 - \dfrac{64}{5}$

 $K = (6)(6) - (4)^2 = 36 - 16 = 20$ and $f_{xx} = 6$ $\hspace{1.1cm}$ $y = \dfrac{-14}{5}$

 $f\left(\dfrac{16}{5}, \dfrac{-14}{5}\right) = \dfrac{-82}{5}$ is a relative minimum.

13a. $R_x = -16x + 39 - 2y$ $\qquad$ $-16x - 2y + 39 = 0$ $\qquad$ $80x + 10y - 195 = 0$

 $R_y = -10y + 39 - 2x$ $\qquad$ $-2x - 10y + 39 = 0$ $\qquad$ $\underline{-2x \ - 10y + \ 39 = 0}$

 $R_{xx} = -16$ $\hspace{6.7cm}$ $78x \hspace{1.5cm} - 156 = 0$

 $R_{yy} = -10$ $\hspace{8cm}$ $x = 2$

 $R_{xy} = -2$ $\hspace{6.8cm}$ $-2(2) - 10y + 39 = 0$

$\hspace{9.5cm}$ $y = 3.5$

 $K = (-16)(-10) - (-2)^2 = 160 - 4 = 156$ and $R_{xx} = -16$

 (maximum)

13b. $R(2, 3.5) = 127.25$ $\qquad$ \$127,250

14. $4x + y - 26 = 0$

 $g(x, y) = 4x + y - 26$

 $F(x, y, \lambda) = 4x^2 + y^2 - 32y + 256 + \lambda(4x + y - 26)$

 $\quad = 4x^2 + y^2 - 32y + 256 + 4x\lambda + y\lambda - 26\lambda$

 $F_x = 8x + 4\lambda = 0$ $\qquad$ $4\lambda = -8x$ $\qquad$ $-2x + 2y - 32 = 0$

 $F_y = 2y - 32 + \lambda = 0$ $\qquad$ so $\lambda = -2x$ $\qquad$ $\underline{-8x - 2y + 52 = 0}$

 $F_\lambda = 4x + y - 26 = 0$ $\qquad$ $2y - 32 - 2x = 0$ $\qquad$ $-10x \hspace{1cm} + 20 = 0$

$$x = 2$$
$$4(2) + y - 26 = 0$$
$f(2, 18) = 20$ is a maximum.
$$y = 18$$

15a, e.

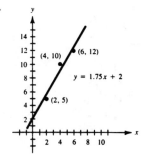

15b. $d_1 = 5 - (2m + b) = 5 - 2m - b$

$d_2 = 10 - (4m + b) = 10 - 4m - b$

$d_3 = 12 - (6m + b) = 12 - 6m - b$

$f(m, b) = (5 - 2m - b)^2 + (10 - 4m - b)^2 + (12 - 6m - b)^2$

15c. $f_m(m, b) = 2(5 - 2m - b)(-2) + 2(10 - 4m - b)(-4) + 2(12 - 6m - b)(-6)$

$\qquad = -20 + 8m + 4b - 80 + 32m + 8b - 144 + 72m + 12b$

$\qquad = -244 + 112m + 24b$

$f_b(m, b) = 2(5 - 2m - b)(-1) + 2(10 - 4m - b)(-1) + 2(12 - 6m - b)(-1)$

$\qquad = -10 + 4m + 2b - 20 + 8m + 2b - 24 + 12m + 2b$

$\qquad = -54 + 24m + 6b$

15d.   Setting $f_m = 0$ and $f_b = 0$:

$112m + 24b = 244 \qquad\qquad -28m - 6b = -61$

$24m + 6b = 54 \qquad\qquad\qquad \underline{24m + 6b = 54}$

$\qquad\qquad\qquad\qquad\qquad\qquad -4m \qquad\quad = -7$

To find b: $24\left(\dfrac{7}{4}\right) + 6b = 54 \qquad\qquad m = \dfrac{7}{4}$

$\qquad\qquad\qquad\qquad 6b = 12$

$\qquad\qquad\qquad\qquad\ b = 2$

$\qquad\qquad y = \dfrac{7}{4}x + 2$

Chapter 9 Test   (pages 538-539)

1. $f(-1, -3) = (-1)^2 - 4(-3)^2 - 2(-1)(-3) = 1 - 36 - 6 = -41$ (b)

2. $d = \sqrt{(-3-1)^2 + (7-4)^2 + (5-5)^2} = \sqrt{16+9+0} = \sqrt{25} = 5$ (d)

3. (d)

4. $\dfrac{\partial f}{\partial y} = -9x^2 y^2 - 7$ (a)

5. $f_x(x,y) = 24x^3 - 6xy^3$

    $f_{xx}(x,y) = 72x^2 - 6y^3$

    $f_{xx}(1,2) = 72(1)^2 - 6(2)^3$

          $= 72 - 48$

          $= 24$ (c)

6.

x	y	x^2	xy
0	4	0	0
-2	0	4	0
1	5	1	5
sum -1	9	5	5

$$m = \frac{3(5)-(-1)(9)}{3(5)-(-1)^2}$$

$$= \frac{24}{14} = \frac{12}{7}$$

$$b = \frac{(9)(5)-(-1)(5)}{3(5)-(-1)^2}$$

$$= \frac{50}{14} = \frac{25}{7}$$

$$y = \tfrac{12}{7}x + \tfrac{25}{7} \quad \text{(a)}$$

7.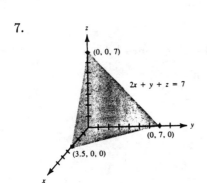

8. center $(2,1,0)$

    radius $\sqrt{9} = 3$

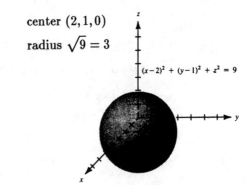

9. $f(0,-1) = 2(0)^4 - 5(0)^2(-1)^3 + 3(-1)^2 = 3$

10. $f_x = 8x^3 - 10xy^3$ 11. $f_y = -15x^2y^2 + 6y$

12. $f_x(0,-1) = 8(0)^3 - 10(0)(-1)^3 = 0$

13. $f_{xy} = -30xy^2$ 14. $f_{yy} = -30x^2y + 6$

15. $f_{xx}(x,y) = 24x^2 - 10y^3$

$f_{xx}(0,-1) = 24(0)^2 - 10(-1)^3 = 10$

16.

$f_x = 3x^2 - 6y - 9$	$6y - 6x = 0$	$3x^2 - 6y - 9 = 0$
$f_y = 6y - 6x$	so $y = x$	$3x^2 - 6x - 9 = 0$
$f_{xx} = 6x$		$3(x^2 - 2x - 3) = 0$
$f_{yy} = 6$		$3(x-3)(x+1) = 0$
$f_{xy} = -6$		$x = 3 \qquad x = -1$
		$y = 3 \qquad y = -1$

Test $(3,3)$: $f_{xx}(3,3) = 6(3) = 18$

$\qquad K = (18)(6) - (-6)^2 = 108 - 36 = 72$ and $f_{xx} = 18$

$\qquad f(3,3) = -25$ is a relative minimum.

Test $(-1,-1)$: $f_{xx}(-1,-1) = 6(-1) = -6$

$\qquad K = (-6)(6) - (-6)^2 = -36 - 36 = -72$

$\qquad (-1,-1,7)$ is a saddle point.

17. $g(x,y) = x + y - 1$

$F(x,y,\lambda) = -3x^2 + 2xy + 4y^2 + \lambda(x + y - 1)$

$\qquad = -3x^2 + 2xy + 4y^2 + x\lambda + y\lambda - \lambda$

$F_x = -6x + 2y + \lambda = 0 \qquad\qquad\qquad\qquad 6x - 2y - \lambda = 0$

$F_y = 2x + 8y + \lambda = 0 \qquad\qquad\qquad\qquad \underline{2x + 8y + \lambda = 0}$

$F_\lambda = x + y - 1 = 0 \qquad\qquad\qquad\qquad\quad 8x + 6y \qquad = 0$

$\qquad\qquad\qquad\qquad\qquad\qquad\qquad\qquad\qquad 6y = -8x$

$\qquad\qquad\qquad\qquad\qquad\qquad\qquad\qquad\qquad y = \frac{-4}{3}x$

$\qquad\qquad\qquad\qquad\qquad\qquad\qquad\qquad x - \frac{4}{3}x - 1 = 0$

$\qquad\qquad\qquad\qquad\qquad\qquad\qquad\qquad\qquad \frac{-1}{3}x = 1$

$\qquad\qquad\qquad\qquad\qquad\qquad\qquad\qquad\qquad x = -3$

$f(-3,4) = 13$ is a maximum. $\qquad\qquad\qquad y = \frac{-4}{3}(-3) = 4$

18a. $R_x = -16x + 136 - 2y$ $80x + 10y = 680$

$R_y = -10y + 36 - 2x$ $\underline{-2x - 10y = -36}$

$R_{xx} = -16$ $78x \qquad\quad = 644$

$R_{yy} = -10$ $x = \dfrac{644}{78} = \dfrac{322}{39} \approx 8.26$

$R_{xy} = -2$ $-2\left(\dfrac{322}{39}\right) - 10y = -36$

$\qquad\qquad\qquad\qquad -10y = -36 + \dfrac{644}{39}$

$\qquad\qquad\qquad\qquad\quad y = \dfrac{760}{390} = \dfrac{76}{39} \approx 1.95$

$K = (-16)(-10) - (-2)^2 = 160 - 4 = 156$ and $R_{xx} = -16$

The revenue is a maximum when $x = 8.26$ (826 items) and $y = 1.95$ (195 items).

18b. $R(8.26, 1.95) \approx \$2596.51$

19a. $d_1 = -3 - (-2m + b) = -3 + 2m - b$

$d_2 = -1 - (-m + b) = -1 + m - b$

$d_3 = 2 - (0 + b) = 2 - b$

$d_4 = 0 - (m + b) = -m - b$

$d_5 = 2 - (2m + b) = 2 - 2m - b$

$f(m, b) = (-3 + 2m - b)^2 + (-1 + m - b)^2 + (2 - b)^2 + (-m - b)^2 + (2 - 2m - b)^2$

19b. $f_m(m, b) = 2(-3 + 2m - b)(2) + 2(-1 + m - b) + 0 + 2(-m - b)(-1)$

$\qquad\qquad + 2(2 - 2m - b)(-2)$

$\qquad\quad = -12 + 8m - 4b - 2 + 2m - 2b + 2m + 2b - 8 + 8m + 4b$

$\qquad\quad = -22 + 20m$

19c. $f_b(m, b) = 2(-3 + 2m - b)(-1) + 2(-1 + m - b)(-1) + 2(2 - b)(-1)$

$\qquad\qquad + 2(-m - b)(-1) + 2(2 - 2m - b)(-1)$

$\qquad\quad = 6 - 4m + 2b + 2 - 2m + 2b - 4 + 2b + 2m + 2b - 4 + 4m + 2b$

$\qquad\quad = 10b$

20.   Setting $f_m = 0$ and $f_b = 0$:

$\qquad -22 + 20m = 0 \qquad\qquad 10b = 0$

$\qquad\qquad\quad m = \dfrac{22}{20} \qquad\qquad\quad b = 0$

$\qquad\qquad\quad m = 1.1$

$\qquad y = 1.1x$